AN ECOLOGY OF HAPPINESS

AN ECOLOGY OF

Happiness

ERIC LAMBIN

Translated by TERESA LAVENDER FAGAN

THE UNIVERSITY OF CHICAGO PRESS

Chicago & London

ERIC LAMBIN is the George and Setsuko Ishiyama Provostial Professor in the School of Earth Sciences at Stanford University and a senior fellow in Stanford's Woods Institute for the Environment, and professor of geography at the University of Louvain, Belgium.

The University of Chicago Press, Chicago 60637
The University of Chicago Press, Ltd., London
© 2012 by the University of Chicago.
All rights reserved. Published 2012.
Printed in the United States of America

21 20 19 18 17 16 15 14 13 12   1 2 3 4 5

ISBN-13: 978-0-226-46667-5 (cloth)
ISBN-10: 0-226-46667-1 (cloth)

Originally published as *Une Écologie du Bonheur*
© Éditions Le Pommier, 2009

Ouvrage publié avec le soutien du Centre national du livre–ministère français chargé de la culture / This work is published with support from the National Center of the Book–French Ministry of Culture.

Library of Congress Cataloging-in-Publication Data

Lambin, Eric F., author.
    [Une écologie du bonheur. English.]
    An ecology of happiness / Eric Lambin ; translated by Teresa Lavender Fagan.
      pages ; cm
    Includes bibliographical references.
    ISBN-13: 978-0-226-46667-5 (hardcover : alkaline paper)
    ISBN-10: 0-226-46667-1 (hardcover : alkaline paper)   1. Human ecology.   2. Human beings—Effect of environment on.   3. Well-being.   4. Happiness.   5. Nature conservation—Social aspects.   6. Nature conservation—Psychological aspects.
7. Environmental psychology.   I. Fagan, Teresa Lavender, translator.   II. Title.
GF51.L3513   2012
304.2—dc23

                                                                            2011043525

♾ This paper meets the requirements of ANSI/NISO Z39.48-1992
(Permanence of Paper).

Only after the last tree has been cut down,

Only after the last river has been poisoned,

Only after the last fish has been caught,

Only then will you find that money cannot be eaten.

CREE INDIAN PROPHECY

CONTENTS

INTRODUCTION

DEGRADATION OF THE ENVIRONMENT, IMPROVEMENT OF WELL-BEING

To begin, this book is based on an observation that today can no longer be contested: in the course of the last century, human beings have profoundly changed planet earth. These changes have modified nature's ability to provide the natural goods and services essential to ensuring the well-being of human societies. We don't need to recall the litany of ecological alterations that are occurring on a global, regional, and local scale; other works, including my earlier book, *The Middle Path*,[1] analyze the nature and the causes of these transformations, as well as possible responses to them. We need only remember that changes to the environment have accelerated considerably since the mid-twentieth century. Human activity has pushed the earth's ecosystem beyond the boundaries of its natural course. Considering the pressures exerted by human beings, the future of our planet has become unforeseeable. Given the complexity of the planet's ecosystem, severe impacts on the environment cannot be ruled out once certain critical thresholds have been crossed, particularly if abrupt variations in the climate and rapid changes in political, economic, or social spheres occur simultaneously, as certain regions of the world are more vulnerable than others to such changes. Humanity has thus involuntarily undertaken an extensive experimental voyage without a clear vision of where it will end up, and without any real captain at the helm.

This rather disturbing statement is, however, countered by somewhat reassuring data that might even give us the impression that environmental degradation in fact has little impact on what is truly important to us. Indeed, during the period in which the environmental changes caused by humans have been the most rapid, there has been a continuous increase in our physical well-being. Life expectancy has gone from 24 years in the year 1000 to a global average age of 66.6 years in 2009 (in the United States it was, on average, 78.3 years in 2005–10). Over the same period, average income has increased by a factor of 20,

---

1. E. Lambin, *The Middle Path: Avoiding Environmental Catastrophe* (Chicago: University of Chicago Press, 2007).

after adjustments allowing for inflation. Since the 1950s, the global economy has been more dynamic than ever. During the second half of the twentieth century, global food production increased at a higher rate than that of population growth. Over an even more recent period, global infant mortality has gone from 200 infant deaths per 1,000 live births in 1980 to 42.1 infant deaths per 1,000 live births (and 6.3 infant deaths per 1,000 in the United States) in 2009; average heights, weights, and IQs have been continuously increasing; cardiovascular, pulmonary, and circulatory illnesses occur much later in life; and chronic illnesses have diminished by 0.7% per year since 1900. Such progress thus suggests that the overall picture of economic development has been largely positive for humanity, in spite of the profound modifications to the natural environment. Globalization can be credited in part with this progress. Since 1950, the growth of global commerce has been twice as rapid as economic growth, which suggests that the increased globalization of trade is intimately associated with the recent period of economic prosperity. Why, then, should we modify the ways we consume or our means of production, when we have been benefiting so greatly from them for more than two centuries now?

The issue is that of knowing whether this increase in human well-being will continue in the decades to come, in spite of the erosion of our natural capital. Will it be possible to extend to the entire human population the progress which up to now has been enjoyed by only the most affluent occupants of the planet—who represent only 20% of the world's population? Don't environmental changes threaten to have negative impacts on human health, safety, and happiness in the decades to come? Will the bill for our very profound modification of planet earth soon be presented to us in the form of a decrease in well-being?

THE IMPORTANT QUESTION

The question that has dominated the debate in the past few years has been this: what is the impact of human activity on the natural environment? This book responds to a complementary question, unasked up to now, although it is crucial if we wish to mobilize all the world's citizens to assist in a successful transition toward sustainable development: what is the impact of environmental changes on human well-being? Or: do we need nature in order to be happy? In more technical terms: is it necessary to maintain the integrity of natural ecosystems to ensure our happiness?

To be clear at the outset, by "environmental changes" I mean much more than global warming: recent phenomena linked to human activity, such as changes in the earth's land ecosystems (through deforestation, land degradation, and agricultural expansion and its intensification) and marine ecosystems (through overfishing, the acidification and pollution of the oceans), a loss of biodiversity, invasions of regions by new animal and plant species, a perturbation of the water cycle, urbanization, and the multiple forms of pollution of the air, water, and ground, are also involved.

The question addressed in this book has an important practical resonance. It deals with an essential motivation to change our mode of development, setting it on a trajectory that is less destructive to nature. Should we decrease our ecological footprint simply to preserve the integrity of nature because it has intrinsic value, or should we do it instead to avoid future catastrophes? Or should we do it because it is essential to increase the well-being of the poorest on the planet—those 3 billion people who live in conditions of severe poverty? Unless our main objective is to maintain or even further increase the level of well-being in the richest countries . . .

Indeed, if there were strictly anthropocentric reasons to decrease our impact on the environment, and even profoundly egocentric motivations—"I am defending my happiness"—it would be easier to ensure everyone's involvement in this vast enterprise, in which we will be engaged in the decades to come. If, however, only purely altruistic motivations justified the adoption of a more sustainable mode of development in the name of responsibility toward future generations, other living species, or, even more abstractly, nature, it would be difficult to mobilize the majority of people. And a revolution in our modes of consumption and production—the first truly global revolution—will be successful only if everyone participates: rich and poor countries, public and private organizations, conservatives and reformers, producers and consumers, actors and spectators, the young and the old. And time is of the essence: to avoid crossing critical thresholds of environmental change, a transition toward a more sustainable mode of development should be carried out between now and approximately 2050. Given the inertia of the globalized economic and natural systems, this objective will be reached only if the entire global community is mobilized, not tomorrow but today.

THE PROBLEM OF "FREE RIDERS"

Although everyone is beginning to see ever more clearly the magnitude of the environmental changes provoked by human activity, changes in behavior and policies remain weak and slow compared with the size of the task. In that respect, individual attitudes within societies are very heterogeneous. A small percentage of the population, less than 20%, is ready to change their mode of consumption for ethical reasons and for the principle of responsibility vis-à-vis nature and future generations. These people wish to cooperate for the common good and are concerned with contributing to the good of the group, beyond their private interests. Some of these people have an "ecocentric" value system—they respect nature for itself—whereas others are "anthropocentric altruists"—they are above all concerned with human well-being, but recognize that this is related to the protection of the environment. This fraction of the population is already convinced, and contributes daily to a transition toward sustainable development, even if it means assuming the costs themselves. These are the people who sort their trash, have already replaced traditional lightbulbs with compact fluorescent bulbs in their homes, use their bikes or public transportation when possible, and instill their lifestyles, their work, and the education of their children with a value system based on responsibility toward others and the world in general.

Alongside those people, the largest proportion of society—around 60% of the population—is made up of "followers" (or "conditional cooperators"), that is, individuals who will follow the movement established by the majority and by opinion makers. They are ready to contribute to the common good as long as everyone else does. They adopt a conditional strategy, which is adapted to the behavior of others. They won't make sacrifices when others are profiting from the system. They follow the movement established by the leaders, but don't want to assume leadership. To make this majority move toward sustainable development thus necessitates first making other segments of society act differently.

But to do this, the problem represented by the fraction of individuals in all societies—the remaining 20%–30%—who are essentially motivated by the pursuit of their own interests without any altruistic concerns, or who are incurable skeptics, must be resolved. These are the "free riders" who use up more than their fair share of resources and assume less than their share of the associated costs. They benefit from the common good without doing much to ensure it is maintained. This group is made up of those who are motivated by the lure of money and by a spirit of competition, who want to be the best and are ready to do anything to achieve their ends. In the environmental realm, these are the people (or leaders of countries) who allow others to decrease their ecological footprint while they happily increase their energy consumption and pollution. For example, there is the Saudi prince who, for a private jet, ordered an Airbus 380, the new jumbo airplane with 840 seats. Each minute of flight he might spend in its onboard gym, or in one of its meeting rooms, will cancel out any effort at energy conservation by thousands of households.

Of course, these free riders represent only a bit more than 20% of the population, but they play a determinant role by the way in which they influence the opinion of followers. A follower will not change his lifestyle if his neighbor is a free rider. Furthermore, these free riders consume so much more than other members of society that they alone contribute to most of the environmental problems. Steven Pacala, an American ecologist at Princeton University, in 2007 calculated that half the world's emissions of carbon dioxide (CO_2), a greenhouse gas responsible for global warming, originate from only 7% of the world's population, or from the 500 million richest people on the planet. By contrast, half the world's poorest population is responsible for only 7% of global carbon dioxide emissions—a negligible amount, which is explained by the poverty and destitution of those 3.3 billion people. The 500 million people who emit the most CO_2 are the free riders in the world system. They are rich relative to the rest of humanity and live mostly in North America and western Europe, but also in Russia, China, and India. They pollute much more than would be their share in an equitable system, but they are also those who decide what the world will be like in the future, owing to their enormous powers in investment and innovation, and to the examples they set vis-à-vis the many people who aspire to become like them. But the greatest philanthropists, those who, after a life devoted to making the most money possible, take it upon themselves to help the poorest—also come from their ranks, even if a few thousand of these very generous donors represent only a small proportion of the 500 million free riders.

Psychologists warn us that an individual cannot be assigned unequivocally to one of the three types described above—altruists, conditional cooperators, and free riders—because personality traits adapt to life circumstances and can change over time. One can behave as a free rider in professional life and be the greatest altruist in his or her community. Nevertheless, recent research in experimental economics has shown that the distribution of these three types within a population is relatively stable throughout the cultures of the modern world. This is revealed through a simple experiment involving public goods, which has been repeated many times worldwide. It is known by the name of Voluntary Contribution Games. In one of the forms of this game, each member of a group of four people selected at random and placed in a laboratory receives a sum of money. Each person can either keep that money by depositing it in a private account, or deposit it in an account for the entire group. The bank doubles the amount deposited in the group account. The rules of the game state that at the end of every round, the amount of money thus obtained is divided into four equal parts and redistributed among all the members of the group, whether or not they deposited money into the group account. Thus, all money deposited into the common account increases the earnings of all the members of the group, but may decrease individual earnings. For example, if the four people each put $10 into the common account, the total would be $40, which would be doubled by the bank ($80), then divided by 4, and each individual would earn $20. If a free rider lets the other three members of the group deposit their $10 into the group account, but keeps his $10 in his private account, then the $30 of the group account will be doubled by the bank ($60) and divided by four ($15 each). The free rider will thus have the $10 in his private account, plus the $15 from the group account, or $25, which represents earnings above the $20 he would have earned by depositing his money into the group account, as the others did. Overall, then, the group as a whole earns less money when one or several of its members adopts the free rider approach. The social dilemma is that the optimum at the collective level is achieved only if each individual makes a decision against his personal desire to maximize his own earnings.

When this game is repeated many times, we discover that more than 20% of subjects behave as free riders. Once the members of a group discover that one of them has adopted that strategy, around 60% of subjects stop depositing their money into the common account—these are the conditional cooperators or followers. Fewer than 20% of subjects, however, continue to invest their money into the common account, regardless of the strategy adopted by the others. These are the altruists. Similar proportions are found for several variations of this game, which is, of course, a caricatural representation of human societies. Real situations and our reactions when we are confronted with them are composed of multiple dimensions, notably affective, which this game cannot represent.

It is nevertheless indispensable to try to understand these different sources of motivation, which are at the foundation of individual approaches, in order to get all members of a society to actively move toward sustainable development. The social dynamics in issues involving the common good always arise from contrasting attitudes among the various components of society. Convincing the free riders to become altruistic is a difficult task. Yet, showing them that it is in their egocentric interest, in order to ensure their happiness, to decrease their ecological footprint is a much more promising approach. They are often innovators endowed with a spirit of enterprise and ready to take risks. As such, they represent the drivers of society. And if, thanks to arguments directed at free riders, the followers become more proactive rather than waiting for others to show them the path, the transition toward sustainable development will be even quicker. As for the altruists, although an argument that connects a decrease in their ecological footprint to their personal well-being isn't necessary, they deserve to be reassured that they have made the right choice by investing in the common good, both for others and for themselves.

Is it possible to find a rational argument, one based on well-established scientific data, that can convince people there is an anthropocentric, indeed egocentric interest in decreasing one's impact on the natural environment? To do this, we must establish a strong relationship between the environment and individual well-being: to become happier, one must protect the environment. Such a positive argument would essentially replace the alarmist discourse—of which many people are tired—thereby accelerating the transition toward sustainable development. The rhetoric of fear, which warns of a collapse of our civilization unless we abandon our current way of life, engenders denial among skeptics, cynicism among nihilists, despair among pessimists, and rejection by optimists.

In this book, I am seeking an approach that will promote the benefits of having a closer relationship with nature and respect for its integrity. I am convinced that an argument that promotes the advantages of a protected natural environment for our happiness, health, and security is likely to convince the greatest number of people, and will encourage attitudes of constructive involvement. It is essential to motivate people to adopt appropriate individual behaviors and to contribute to collective decisions that respond to the challenges of the twenty-first century. To bring about this change in attitude, we must appeal to a personal source of positive motivation. And what is more important to us than our happiness? Who doesn't wish to improve his or her well-being, health, and feeling of security, and the general satisfaction that he or she derives out of life? Several psychological studies suggest that the happier people are, the more they are inclined to adopt a respectful attitude toward the environment. And if a less degraded natural environment will make us happier, we would then enter into a virtuous, mutually reinforcing cycle of a conservation of nature and an increase in personal happiness.

WELL-BEING AND THE NATURAL ENVIRONMENT

In the last few years, economists have discovered—finally! some will say—that economic growth does not automatically lead to an increase in the happiness of the population, nor does it always maintain the flow of goods and services provided by nature. Words such as *happiness, well-being, satisfaction,* or (in economic terms) *utility* describe with some important nuances the response that each individual might bring to a question such as: all things considered, how are things going for you these days? Would you say that you are very happy, happy enough, or not very happy? Thus, we are looking at an individual's subjective evaluation of his or her situation over all the dimensions that affect his or her existence and over a sufficiently long scale of time to eliminate the effect of small contingent events.

Many studies have identified the various factors that contribute to a happy existence. In general, a happy, quality life is associated with the existence of opportunities, with the meaning and the goals people assign to their lives, and with the ability to enjoy the possibilities we are all offered. There are five categories of factors that determine a happy life.

1. Personal situation: health, affective life, leisure, work, mobility . . .
2. A feeling of security: the fear of criminality, conflicts, wars . . .
3. The social environment: belonging to a network of relationships, feeling of confidence, availability of help in case of need . . .
4. The institutional environment: freedom, political involvement, the proper functioning of the justice system . . .
5. The natural environment: exposure to noise and pollution, access to preserved natural spaces, the feeling of being connected to nature . . .

Some of these factors—for example, family life, financial situation, the interest of professional work, insertion in a community and network of friends, individual freedom—do not depend on the natural environment. Other factors depend directly or indirectly on the natural environment, notably health and one's degree of vulnerability in the face of infectious and noninfectious diseases, physical well-being related to access to clean water and to food in sufficient quantity and quality, and personal security connected to natural catastrophes and conflicts that could originate in a control over natural resources or climate change. All dimensions of human existence occurring within the natural framework of life are also part of this category, including the environment of the place where one lives, outdoor leisure activities, contact with domestic or wild animals, and the spiritual, aesthetic, and symbolic life connected to the natural world. Finally, personal values that include a pursuit of the common good are also connected to the natural environment, which inspires a feeling of responsibility toward others, the world, and future generations.

This book explores the way in which these various dimensions of human happiness are affected by changes in our natural environment, and the impact those changes have on human well-being. The term *ecology* (from the Greek *oikos*, "house," "habitat," and *logos*, "study of") was introduced in 1866 by German biologist Ernst Haeckel. He defined ecology as "the study of the relationship of organisms with their environment, that is, in the broader sense, the study of conditions for existence." An ecology of happiness is thus an attempt to understand the interaction between human happiness and our environment: in what environmental conditions can we experience happiness?

This book's thesis is that humans have an interest in preserving nature, because our happiness depends greatly on the natural environment. Three components of well-being are examined: the subjective perception of a happy existence, health, and security. The impact of environmental changes on physical well-being (including food production, access to good-quality fresh water, the use of renewable and nonrenewable natural resources) is not looked at here, because that more general question has already been dealt with competently in many other works.

The initial chapters reveal all that we have to gain from reconciling with nature. Recent research in the social sciences on the many psychological and health benefits of a close relationship with nature is looked at in chapter 1, and the various ways we interact with the animal world are examined in chapter 2. The chapters that follow deal with the health and security risks connected to a profound and rapid change in the natural environment. They look at the impact of environmental changes on the emergence of infectious diseases (chapter 3) and on illnesses transmitted by vectors (chapter 4); the health aspects of globalization (chapter 5); the impact of urbanization on health (chapter 6); the issue of conflicts (chapter 7); and the displacement of populations in relation to environmental changes (chapter 8). I conclude with a presentation of the innovative policies of three developing countries, Vietnam, Costa Rica, and Bhutan, that have set goals to better reconcile their development with the environment (chapter 9). In the conclusion, I identify some choices indispensable to our mode of development as well as win-win strategies that might engender benefits for both the environment and human well-being.

In a letter Charles Darwin wrote, "On the other hand I sometimes think that general and popular treatises are almost as important for the progress of science as original work."[2] This book offers a synthesis of recent original research (references for which are provided by relevant chapter in the bibliography) from very different specialized areas, and which are rarely examined alongside one other. And yet, all these areas of research deal more or less directly with various dimensions of happiness. Such rather heteroclite assembling is not encouraged by the structure of our scientific institutions, which are still often organized around disciplines, not social issues. Nevertheless, a coherent image emerges from this synthesis, which once again validates the principle that the whole is worth more

2. From F. Darwin, ed., *The Life and Letters of Charles Darwin* (Teddington, UK: Echo Library, 2007), 2:489.

than the sum of its parts. In my investigation into the ecology of happiness, I present only propositions that have been subjects of empirical verification, following rigorous scientific criteria.

SEPARATION

In this attempt to understand the impact of environmental changes on human happiness, we find two recurrent themes: the separation between rich and poor countries and, in the rich countries, the separation between human beings and nature. Regarding the first theme, an essential aspect of the relationship between happiness and the natural environment concerns the way in which both the progress in increasing well-being and the negative impacts of the degradation of the environment are distributed among the different regions of the world, in particular among rich and poor countries and, within those countries, among the rich and the poor. Of course, there has been an increase in well-being during the last few centuries, but then enormous social inequalities become all the more unacceptable. Half the world's population still lives in great poverty. The income gap between the richest countries (20% of all countries) and the poorest countries (also 20%) has a factor of 30. More than 80% of the world's population lives in countries where the gap between the income of the richest and that of the poorest has been increasing since the 1990s. The gross national product of Belgium (the wealth it earns each year) is equivalent to the gross national product of all of sub-Saharan Africa—which has 770 million inhabitants compared to 10 million in Belgium. In developing countries, low body weight is the primary risk factor for illnesses, whereas in high-income countries being overweight is one of five primary risk factors.

In the short and midterm, the negative effects of environmental changes on well-being will be felt essentially in the poorest regions of the world, which are also the most vulnerable. The consequences of environmental changes on happiness thus risk the further increase of social inequalities on a global scale, via their effects on health, security, and quality of life. The most pernicious aspect of human activity's impact on the earth's ecosystem is no doubt more social than ecological or climatic. The mechanisms used by rich countries to protect themselves against the most negative impacts of environmental changes contribute to increasing inequalities in development among regions of the world. The solution to environmental problems thus clearly calls for an anthropocentric motivation. For populations of the richest countries, this motivation is based on a moral imperative of solidarity vis-à-vis the most destitute portion of humanity.

Furthermore, the strategies that the affluent of the planet implement to protect against climate change, a decrease in biodiversity, pollution, and a modification of earthly ecosystems and the water cycle result in separating individuals from nature even more. Sheltered in our cities, secure in our cars on our paved roads or in our heated and air-conditioned houses, we are detached from what is at the heart of humanity: our biological roots, which plunge deep into the natural world; our psychic relationship with the diversity of life forms; an anchoring in the beautiful landscapes; and a fraternity with the animal world. Our existence henceforth unfolds in a world of artifacts that evoke the technological superiority of the human species while omitting the importance of the role played by nature. The environmental changes for which we are responsible come at a high price, in terms of both health and happiness. An appeal to an egocentric motivation—ensuring our individual happiness through contact with a preserved nature—to resolve environmental problems is aimed above all at the most affluent populations of the planet.

THE EXPERIENCE OF NATURE

At the heart of what I call the ecology of happiness is this question: how does a close and daily contact with nature influence one's subjective perception of happiness? In other words, is human life enriched by experiences with the natural world? Can one live happily without an intimate relationship with it? Where does the satisfaction we derive from looking at natural settings come from? What makes us happy in our relationships with the animal and even the plant kingdom (an example: the pleasure of gardening)? Does the degradation of the environment irreparably lead to an impoverishment of the human experience and a lack of happiness? Can we live happily while being separated from the natural world by a screen of material artifacts?

DISAPPOINTING MATERIALISM

The most recent studies on happiness, whether in economics, psychology, or sociology, suggest that beyond a threshold of basic material comfort, money does not buy happiness, nor can the possession of material goods increase or sustain the satisfaction one derives out of life. This conclusion has been reached in various studies, all rigorously reproduced in many cultural contexts, over various age groups and social classes. American sociologist Tim Kasser has shown that people whose value system is highly materialistic suffer from lesser personal well-being and from worse mental health than people who are not very materialistic. Regardless of the way in which materialistic values and well-being are measured, the same conclusion is reached: adopting materialistic values is both a symptom of a sense of insecurity and an inappropriate strategy for easing a sense of dissatisfaction. Materialism has negative consequences on emotional well-being and increases anxiety. The relentless pursuit of an accumulation of material goods leaves very little time to devote oneself to that which truly creates happiness—for example, family, friends, one's community, a truly gratifying job, and leisure activities that have a positive effect on one's physical and mental health. The satisfaction of materialistic desires also forces one to compromise personal freedom and value systems.

As regards individual countries, all changes that increase citizens' free choice—democratization, greater social tolerance of diverse lifestyles, and economic growth—are associated with an increase in happiness for most of the members of society, on the condition that the benefits of that development are widely shared within the society, and that increased opportunities are offered to everyone. The relationship over time between the happiness of individuals and their income is weak. Moreover, for poor countries, economic growth translates to an elevation in happiness for the population, but beyond a gross domestic product (GDP) per capita of $12,000–$15,000 per year, the increase in happiness that accompanies economic growth is smaller, or even nonexistent in a few cases (in 2009, the United States' GDP was at $47,000 per capita, at purchasing power parity). For example, an increase of $100/year in average income will have an impact on well-being that is twenty times higher in a very poor African country than in the United States. Beyond a certain threshold of prosperity, the noneconomic dimensions of human existence become preeminent in the definition of quality of life.

American economists Daniel Kahneman, Nobel Laureate in Economics in 2002, and Richard A. Easterlin have both shown that when someone's income increases, his material aspirations increase just as rapidly. New aspirations (for a bigger car, a vacation home, a flat-screen TV) are created as soon as earlier desires have been satisfied. Over a period of close to forty years, the Gallup Poll asked the following question to a large sampling of Americans: what is the income needed by a family of four to live comfortably in your community? In response to this question, year after year most people calculated an amount that increased as rapidly as their true income. There is thus a "hedonic adaptation" to a person's new pecuniary circumstances: the level of well-being remained unchanged or increased only temporarily after new goods were acquired. This was less true for nonmaterialistic aspirations related to social, family, and cultural life, for which this adaptation was incomplete. Denmark is consistently one of the happiest countries in national surveys on happiness. Its secret seems to be in the modesty of its citizens' expectations. To be happy, is it enough to decrease one's desires rather than increase one's income?

Each person's level of satisfaction is also defined through social comparisons: the pleasure we derive from our material possessions depends in part on the amount of those same goods that others possess. Indeed, many studies have shown that a person's ranking on the income scale within a population, that is, his or her relative assets, affects his or her degree of satisfaction much more than assets measured in absolute terms. To be among the richest in one's reference group is more important than achieving a particular level of income. As a corollary, those who adopt a new reference group of people who are more affluent than they are discover they are less happy, even when their income is stable or increasing. If the entire population becomes richer at the same rate, the increase in each person's happiness is small. For example, the spectacular economic growth of the United States between 1946 and today has not been accompanied by an increase in the number of people who consider themselves very happy—the subjective happiness among American women even declined during that period. This fact disproves one of the fundamental postulates of most macroeconomic analyses; that is, that the average income of a country is a good measure of the well-being of its population, and that a rapid growth in the GDP should be a primary objective. When the income of people increases, they have a tendency to allocate a higher proportion of their time to activities that not only elicit little satisfaction but, on the contrary, increase stress and pressure: a demanding job, long daily commutes, pointless shopping, restricted leisure activities.

Thus, once a threshold of prosperity has been crossed, money and material goods contribute little to happiness. Generally, everything related to material possessions makes us less happy than that which derives from actual experiences. To engage in activities such as hiking in the woods or reading not only makes us happier, but also has a much smaller impact on the environment than an acquisition of material goods, whose production and use consume energy and materials and produce waste.

If we are made aware that engaging with nature truly makes us happier, then a virtuous cycle will be set in motion. Because the pursuit of a happy life will then be associated with a mode of consumption whose environmental impact will be diminished, which will preserve nature and will thus increase opportunities to increase one's happiness in contact with it. But has this hypothesis been verified?

LET THE STATISTICS SPEAK

Statistical data collected at the level of both countries and individuals have enabled us to determine whether people who say they are the happiest live in regions whose natural environment is the best preserved. In particular, there are a few statistical studies led by economists such as John M. Gowdy, Heinz Welsch, and their colleagues at the beginning of the 2000s. These studies are based on surveys pertaining to life satisfaction. Each year several countries conduct a survey over a large random sampling of citizens, asking a question such as: "On the whole, would you say you are: very happy; quite happy; not very happy; not at all happy?" Some surveys focus on feelings of happiness, whereas others look at life satisfaction, which introduces a slight nuance. A complementary question is often: "Are you satisfied in specific areas of your life: your job? Your financial situation? Your place of residence? Your health? Your leisure time? Your environment?" Respondents then have the choice between a limited number of responses, such as "not satisfied," "satisfied," "very satisfied." These surveys provide a subjective measurement of well-being, whose validity has been established by a large number of experimental studies in psychology and neurobiology. They have been conducted at regular intervals in several countries beginning in the late 1940s, and in a steadily increasing number of countries in the 1970s and 1980s.

One of the first statistical analyses to explore the relationship between the results of these surveys and environmental factors involved 20,000 German citizens in 2003. It showed that the overall satisfaction that individuals derived from their lives was explained, in order of importance, by satisfaction related to their financial situation, their health, and their job, with the environment being considered an unimportant factor. However, a significant portion of the variations in individuals' well-being was linked to single factors that were not measured in this study. In Holland, it was shown that people who live near an airport and endure the sound pollution from it are generally less satisfied with their lives than others. Another survey showed that living in the heart of a large city such as London also decreases well-being. A much more complete study conducted in Ireland, published at the end of 2008, showed that individual happiness is linked not only to socioeconomic and sociodemographic factors but also to environmental factors. As in other countries, the happiness of the Irish is associated with having a job, a good level of education, and a good income (up to a certain threshold of income, however), along with being in good health and not being divorced (women are generally happier than men in Ireland). Furthermore, happiness increases with factors involving the climate (high temperatures, mild winds, and a level of rainfall associated with green landscapes), when the density of the population is relatively high (which is reasonable in that rather sparsely populated country) and when one's home is a few kilometers from the sea. By contrast, living near a source of pollution or a busy highway not surprisingly decreases individual satisfaction significantly.

Other statistical analyses have been carried out on the national rather than the individual level. In these, each country is represented by the average response of its citizens to surveys on subjective satisfaction as well as by a series of socioeconomic and environmental factors. An analysis based on a few European countries determined that the level of air pollution relates significantly to differences in subjective perception of well-being, whether over different countries or over time: people who declared that they had a high level of well-being more frequently lived in countries where the air quality is better. Another study carried out over sixty-seven countries has shown that the climate—particularly where temperatures are highest or lowest—strongly influences perceptions of degrees of happiness, independently of factors such as per capita income, unemployment rate, political freedoms, life expectancy, or density of the population; subjective satisfaction increases in countries where the average temperature of the coldest months is higher (milder winters), and it decreases in countries where the average temperature of the hottest months is higher (torrid summers). It also decreases in climates characterized by several months with very low precipitation (long dry seasons). This implies that with global warming, people will become more satisfied with their climate conditions in high latitudes, where the richest countries are found, and less happy in equatorial and tropical regions, that is, in the regions of the world where most of the poorest countries are concentrated. The conclusions of this study, based on the projection of a simple statistical relationship, nevertheless reflect an environmental determinism that is, however, widely discredited by the facts: societies adapt their infrastructures and their ways of life to the climate, and the equatorial and tropical regions are filled with very happy people.

Other studies have analyzed the connection between well-being and the culturally connected attitudes of individuals vis-à-vis the environment. Based on a sampling of 9,000 British citizens, a detailed statistical analysis carried out in 2007 has shown that the more people are concerned with pollution and the degradation of the environment, the destruction of the ozone layer given as an example, the more they consider their well-being to be mediocre. On the other hand, the more they are involved with the living world and the preservation of biodiversity, the higher their perception of their well-being. The authors of this study interpreted these results in the following way: an expressed concern involving a negative aspect of the environment (its pollution) is associated with a decrease in a feeling of happiness, whereas a psychological relationship with living beings and other species, which reflects a positive view of the environment, is associated with an increase in a feeling of satisfaction. The analysis was carried out in such a way that the results indeed reflected the effect of a profound concern for the environment and not only the fact that some individuals endure a higher degree of pollution (which indeed decreases their well-being), whether they practice more leisure activities in nature (which makes them happier), or whether they have a rather optimistic or pessimistic psychological inclination toward life in general: through a series of well-conceived questionnaires, the effect of environmental attitudes was isolated from these other factors.

These statistical results thus show that there is indeed a relationship between happiness and the natural environment. They also suggest that policies aiming to resolve problems of pollution and to preserve biodiversity have a positive impact on human well-being.

THE VIRTUOUS QUARTET

On the scale of countries, another study, published in 2007 by Slovenian physicist Aleksander Zidanšek, found a positive correlation between the subjective satisfaction of a country's inhabitants regarding their lives and an indicator that measures the environmental performances of that country. This standard indicator was determined for each country from measurements such as water and air quality, the amount of protected zones, greenhouse gas emissions per inhabitant and per unit of GNP, and so forth. The countries in which the inhabitants are the happiest are also those that implement the most sustainable development policies with regard to the environment.

But is it sustainable development that makes people happy, or is it the happiness of people that causes them to develop a larger sense of responsibility toward nature and future generations, and thus causes them to decrease their environmental impact? Most likely, a bit of both.

To an even greater extent, two factors underlie both a happy society and a sustainable economy: a good educational system and a "postmaterialistic" value system, that is, values that are less focused on economic and physical security, and more concerned with issues of identity, the pursuit of quality of life, and access to information and knowledge. Surveys on satisfaction with life have shown that people who have a high level of education are in general happier than others. These same people are also more inclined to adopt sustainable modes of consumption. Moreover, people who have a nonmaterialistic value system are clearly happier on average than those for whom money and the possession of material goods are very important.

On the collective level, the transition from materialism toward postmaterialism is also the most important and most fundamental step for the successful passage toward sustainable development. The preservation of nature and an increase in happiness are thus two positive consequences of a profound transformation of society and individuals' value systems. Education, a rich value system, individual happiness, and sustainable development form a virtuous quartet. One of the implications of the correlation between happiness and sustainable development is that policies promoting values and behaviors associated with a smaller environmental impact for economic activity to the benefit of future generations will in addition make the current generation happier!

BIOPHILIA

To develop a theory of an ecology of happiness, we must go beyond these statistical correlations and understand what, in their contact with a preserved natural environment, makes people happy. Looking back at the origins of human beings' relationship with nature is an obvious first step. American biologist Edward O. Wilson, professor at Harvard University and a great pioneer in biodiversity studies, in 1984 proposed the concept of biophilia, from the ancient Greek bios—"life"—and philos—"beloved" or "friend." According to Wilson, people have an innate tendency to establish a relationship with the living world and natural processes. In other words, the human species has an innate emotional affinity with other living beings as well as with the plant kingdom and natural surroundings. This concept of biophilia thus refers to the psychological well-being that people experience during a close interaction with the natural environment. An attraction to nature is the expression of a biological need that has been an integral part of the development of the human species since its origin and which is essential to both the physical and the psychic parts of human nature. The hypothesis of a human dependency on nature implies much more than a need to satisfy one's physical wants; it also includes a search in nature for aesthetic, emotional, cognitive, and spiritual satisfactions, and, more widely, a quest for the meaning of life.

Of course, we all feel an attraction to nature and pleasure in contemplating it, wandering in it, or playing a sport in it. But the concept of biophilia contains an even more fundamental hypothesis: it asserts that this need for an intimate association with the natural environment and in particular with the living world has a biological basis, that it is innate rather than acquired. According to this hypothesis, this attraction is the result of the evolution of our species and has provided humans with a competitive advantage in natural selection over millennia. In humans, genes and culture have coevolved to produce positive adaptive responses to everything in nature that is essential for survival. Ever since the human species appeared, it has spent most of its existence integrated in a natural environment. Urban life and modern technology, which have partially disconnected humans from nature, appeared only very recently. Before that, individuals who had the greatest ability to "read nature," that is, to find potable water, to identify edible plants, to follow the tracks of animals, and to find natural refuges sheltered from danger, had a great advantage in the struggle for survival. Based in part on these abilities, for the 5 million years that have passed since the appearance of hominids, natural selection is still very probably part of the genetic inheritance of modern humans, even among individuals who have lived in cities for countless generations.

Before modern civilization, for around 350,000 generations the human species as we know it today lived, in nature, off of hunting and gathering. Humans may thus still be influenced by the lessons that were part of their survival strategies during the prehistoric era. This, according to Wilson, is what establishes biophilia. This also explains the feeling that a stroll in the forest or elsewhere in a natural setting is in some way a return home; it engenders a sense of going back to one's roots and of renewal. Divorced from nature, modern urban life necessarily has negative psychological consequences, since it is foreign to our long history of a natural existence which has left its genetic imprint, even among the most resolutely urban of humans.

Wilson and other eminent researchers who, after him, have reflected on the concept of biophilia illustrate the biophilic tendency notably through the almost universal human preference for open landscapes, with grassy carpets sprinkled with groves of trees and some bodies of water. These same aesthetic principles of landscape composition are also found in urban parks, golf courses, beautiful cemeteries, large gardens, or works of landscape painters. When several landscape photographs are shown to individuals, their most positive reactions occur before views of savannahs that offer a wide perspective and contain several groves of trees to break the monotony. The same response is given regardless of the culture or the geographic origin of the subject.

Human prehistory quite probably originated in the African savannahs, in a landscape of prairies with very few trees, except along rivers and lakes. For our ancestors, places that were slightly raised, with a few trees where they could hide and a wide view over the plains, offered an ideal surveillance point both to hide from predators and to seek out prey. The presence of water sources enabled them not only to quench their thirst but also to hunt the animals that came to drink. A landscape that was too open diminished the opportunities for hiding; a dense forest increased the risk of a surprise attack by a predator. The most favorable habitats for the survival of our ancestors thus had the same fundamental characteristics as the landscapes that are now considered the most pleasant by modern humans. Today, contemplating those landscapes engenders a feeling of peace, tranquility, and relaxation, as if we are sensing that the conditions for our survival are guaranteed through that natural environment.

This example of savannah landscapes can be extended to other types of landscapes, as well as to biological organisms, both plant and animal, among which the human species has lived throughout its long history. In addition to this "biophilic" response, genetic and cultural evolution has also produced negative responses vis-à-vis dangerous natural elements. Thus, a fear of snakes would be an innate "biophobic" response to the objective danger that venomous snakes represent. Chimpanzees raised in captivity, and which have thus not been exposed to the danger of snakes, nevertheless manifest a very great terror at their first encounter with them, which indeed suggests the innate nature of that fear. Furthermore, the traditional literature and folklore of Ireland, one of the rare countries in the world that have no snakes, contains many references to a fear of snakes, so large a part does that fear play in our biological history.

Anthropologists have also noted that peoples of traditional, nonliterate societies from many different parts of the world have developed categorization and nomenclature systems for plants and animals that display widespread regularities. These observed typological regularities are studied in the field of ethnobiology—that is, the scientific study of dynamic relationships among peoples, biota, and environments. Scholars explain these structural commonalities between classification systems by the similar perception and appreciation of the natural affinities among groupings of plants and animals in their environment—an appreciation that is thought to be largely unconscious. It results from the deep relationship with the natural world that people from traditional societies have established for millennia. Note, however, that the case for the existence of pan-environmental perceptions should not be overstated as there are indeed many cross-cultural differences and nuances.

One of the consequences of the hypothesis concerning the biophilic tendencies of humans is that a profound relationship with natural things is a source of a self-realization and of happiness, since that association is inscribed in human nature. This is why strolling in the woods or spending vacations on a remote beach or in the mountains is so appealing. By contrast, a human being for whom a relationship with the natural world has been highly degraded runs an increased risk of a stifled existence. Thus, according to this hypothesis, preserving the natural world and its diversity is in the profound interest of individuals and humanity.

Stephen Kellert, professor of environmental sciences at Yale University, has identified nine possible dimensions to the biophilic tendency of humans, each of which is embedded in our biological makeup. In a certain sense, these are the fundamental aspects or universal and functional expressions that underlie our affiliation with nature. These nine perspectives all contribute to reinforcing the notion that a conservation of nature procures an evolutionary advantage to the human species. I will summarize Kellert's work here, since it is central to the argument developed in this chapter. These dimensions are identified as follows:

1. Utilitarian

Nature is of material value to humans, because it provides them with essential physical elements for their survival, protection, and safety: water, food, medical products, fuel, tools . . .

2. Naturalistic

A simple, direct contact with nature in different forms (landscapes, animals, flowers . . .) leads to mental and physical satisfaction. In modern society, many recreational and leisure activities exploit that inclination: hiking, river kayaking, rock climbing, horseback riding . . . Beyond a feeling of appreciation, experiencing the complexity and diversity of nature fascinates and delights through its mysterious nature. In human history, this fascination and curiosity vis-à-vis nature have constituted a stimulus for an exploration of it.

3. Ecological-Scientific

Ecological science aims to study nature and its functioning in a systematic and precise way. It is based on the conviction that nature can be understood through empirical investigation based on observation and analysis. Ecological science is less reductionistic than other scientific disciplines, because it bears on the study of interactions between living beings and their physical environment, which form an ecosystem. This science reveals a complexity in the natural world that is invisible to the uninitiated. It encourages a better understanding of natural processes, thus enabling human societies to adapt better to natural constraints. It also opens the path to an increased control of nature.

4. Aesthetic

In all cultures, nature is central to artistic expression. In particular, the representation of what is most majestic in nature—tall, snow-covered mountains;

the largest and noblest animal species, such as the wolf, the whale, or the horse—inspires a universal aesthetic appreciation. World literature is rich in descriptions of natural scenes, from the very brief Japanese haiku to the fluid, flowing novel of Marcel Proust and the eclectic *Journal* of Henry David Thoreau. Nature's preeminence in art reveals its role as the model of harmony, symmetry, and order that humans have bestowed upon it. This aesthetic dimension inspires a feeling of tranquility, peace, and well-being essential to the success of human enterprises.

5. Symbolic

Natural elements are widely used in human communication and thinking, according to an idea that the anthropologist Claude Lévi-Strauss had already proposed in 1970. The diversity of nature suggests notions of category and differentiation, central to the structuring of language. Animals, for example, are omnipresent in myths, tales, legends, and children's stories, in every culture in the world.

6. Humanistic

The humanistic experience of the natural world is characterized by a strong affection for the individual elements in nature, typically for the mammals we frequent. The profound emotional attachment people can have for some of them indicates a certain humanization of nature. There are people for whom houseplants or a flowerbed are essential to their psychic balance.

7. Moralistic

In certain circumstances, people have a feeling of affinity, of moral responsibility, and even of reverence vis-à-vis nature. This is the result of the conviction that the natural world is fundamentally ordered, the bearer of meaning and harmony. Thus, there is a profound relationship between nature, on the one hand, and morality and spirituality, on the other. These feelings are at the heart not only of poetry, religions, or philosophy, but also of science—which is revealed, for example, by Einstein's famous "God does not play dice."[1] Better than other religions, pantheist religions, very widespread among indigenous populations, have expressed this connection explicitly, and it is characterized by a respect for all of nature.

1. A. Einstein, letter to Max Born (December 4, 1926), in *The Born-Einstein Letters*, trans. I. Born (New York: Walker, 1971).

8. Dominative

Mastery of the natural world is at the heart of human experience. To dominate the elements is a necessity, both to protect oneself from unforeseen natural events and catastrophes (dams against floods, firebreaks against forest fires) and to increase the goods that nature provides to humans (fertilizer to increase agricultural production, well-managed forests to obtain high-quality timber). Recent history has taught us, however, that the line between a mastery and a degradation of nature can be easily crossed.

9. Negativistic

Elements in nature can also incite feelings of fear, aversion, and antipathy—and rightfully so. To avoid and flee certain dangers inherent in the natural world have always been and still are factors of survival, whether they involve dangerous animals, venomous plants, or dangerous settings such as mountain glaciers or coastlines exposed to tsunamis. Unfortunately, the reaction to these threats has often been excessive, leading to the extermination of certain animal populations (the wolf, for example).

PSYCHOLOGY OF THE ENVIRONMENT

John Muir, a late nineteenth-century naturalist and writer from Scotland, wrote these wonderful lines:

> Climb the mountains and get their good tidings. Nature's peace will flow into you as sunshine flows into trees. The winds will blow their own freshness into you, and the storms their energy, while cares will drop off like autumn leaves.[2]

Many other poets and artists have expressed the benefits of nature for the body, mind, and soul. They speak of feelings that everyone has the opportunity to experience in the course of their life.

Psychologists have been interested in the influence of the natural environment on feelings, behaviors, thoughts, emotions, and actions. Modern life and its frenetic rhythm often bring about a fatigue in directed attention, that is, attention centered on specific tasks, such as at one's workplace. The symptoms of this fatigue are difficulty concentrating, a tendency to be irritable, and a heightened frequency of errors in tasks that require concentration. This decrease in

2. Found at http://www.quotegarden.com/nature.html—Trans.

cognitive resources in the face of the demands of daily life is a major cause of stress. Many studies have shown that contact with nature is a very effective way to recover from this mental fatigue. The natural world helps us recharge our batteries. Most people perceive natural environments as being more favorable to such revitalization than urban environments. A contact with nature enables the mechanisms on which directed attention depends to rest and thus be restored to their normal level. Recent studies have shown that getting out in nature and moving—"green exercise"—for as little as five minutes improves both mood and self-esteem. Exercising near a body of water has the greatest effect. Natural environments do provide an important health service.

Another important discovery of environmental psychology: the ecological preferences of individuals are influenced by their need for places in which they can find renewal. In other words, if we perceive natural environments as being more beautiful and more attractive than urban environments, it is in part because they respond better to our search for favorable frameworks in which to renew our energy. Many psychological experiments unambiguously confirm that a proximity to nature has beneficial effects on both psychic and physical health. For example, residents of urban neighborhoods whose surroundings are depressing and without natural vegetation seem to suffer more frequently from symptoms of chronic stress and other health problems, independent of characteristics such as age, social milieu, or lifestyle.

In 1984, Roger Ulrich, an American geographer, published a surprising study in the journal *Science*. In it he showed that patients who had undergone a surgical procedure recovered better when they had a hospital room with a view onto a natural setting rather than onto a brick wall. Those with a view onto nature were able to leave the hospital on average one day earlier than those that didn't, they needed fewer painkillers, and their attitude was considered more positive by the nurses. Since then, the results of this pioneering study have not only been confirmed but also expanded to other situations, in particular by the team of Terry Hartig, an American psychologist at the University of Uppsala. A proximity to nature or simply a view of it increases well-being in the workplace. Among people who have a sedentary job, those who have a window with a view onto trees, bushes, or flowers express a greater satisfaction than those whose window looks out over a parking lot, a street, or other buildings. In addition, those latter people suffered more frequently from headaches. Taking short breaks from work to contemplate a natural setting decreases mental fatigue. Plants in an office also have a relaxing effect. In a Michigan prison, inmates whose cell window looked out onto an inner courtyard consulted the medical services over 24% more often than prisoners whose window looked onto a country landscape.

Regarding one's place of residence, the most important source of satisfaction for a sampling of Americans in a residential community close to Detroit was the presence of accessible wooded zones in the immediate environs of their house. This factor was more important than the size of their yard. The paradox is that the creation of peri-urban neighborhoods in forested zones results in the destruction of the forests, whereas they were the main reason the residents were attracted to those areas.

In the short term (for around three weeks after returning), vacations devoted to hiking in nature engender a greater feeling of happiness and better performances on tests that required a great deal of concentration than vacations devoted to visiting friends or family members, a car trip in a nonnatural environment, a cultural trip, or relaxing activities at home. Frequent gardening also enables people to face stress better (doctors even speak of "horticultural therapy").

An image of nature is less efficacious, but it, too, has an appeasing effect: when one of the walls of a dentist's waiting room is decorated with a representation of an open natural landscape, patients have a lower heart rate and say they are less anxious at the time of their appointment. After taking a test on a difficult subject or seeing a film of an industrial accident, students who were shown images of natural landscapes recovered more quickly emotionally than those who were shown urban scenes.

All these results come from rigorous testing in the discipline of experimental psychology, in the course of which many factors likely to bias observations were controlled by highly sophisticated procedures. This research in environmental psychology proves that being in contact with nature, or contemplating it, either in reality or in an image, has a beneficial effect on one's well-being.

EXPLICATIVE THEORIES

In psychology, three theories explain these empirical observations.

According to the first theory, which is related to restoring directed attention, and which is proposed by Stephen and Rachel Kaplan, two American pioneers in environmental psychology, we recover from fatigue caused by directed attention through involuntary attention, which requires no effort. Natural environments in particular lend themselves to that restorative experience, for they have many favorable properties; they create a psychological distance from usual mental preoccupations and thus distance us from routine. Mountains, the sea, and forests offer this feeling of distance. This involves a conceptual rather than a physical transformation or distancing. Through their aesthetic qualities, natural environments foster a "soft fascination" by soliciting the senses without the need for any particular effort, thanks to an almost involuntary attention based on interest and curiosity. When this attention is engaged, solicitations for voluntary attention decrease, which enables us to recover. Such is the power of a sunset, autumn colors in a forest every year, or the ever-changing shape of clouds: these are fascinating without taxing the mind, as reading a text or participating in a meeting would do.

Natural environments also offer an organized and coherent framework of great scope and dimensions so that they create the feeling that there is always more to discover. They thus constitute an ideal support for a continuous exploration to engage the mind. This is the case, for example, of a path whose curves hide new vistas, or of archeological ruins and vestiges that connect us to ancient times, and thus to a wider world than that of our day-to-day existence. Finally, there is a higher degree of compatibility between the inclinations of people and the characteristics of nature, both in what it has to offer, that is, possibilities for exploration, and by what it demands, that is, constraints associated with that exploration. Whether one approaches nature with the objective of moving around in it (hiking, horseback riding, or bicycling) or acting as a predator in it (hunting or fishing), domesticating it (gardening), observing it (birding, flowers), or having adventures in it (scouting, mountain climbing, camping), these diverse aims always encounter one of the multiple facets of the natural environment. This polyvalence of relationships between people and nature has been demonstrated by Kellert's nine perspectives on nature presented above.

Another theory that complements the first accentuates the physiological and emotional changes that can occur in an individual when he or she contemplates a natural scene immediately after experiencing a situation that involved a challenge or a threat. In the early 1980s, Roger Ulrich, who was interested in the architectural concepts behind healthcare facilities, intuited that a perception of certain contents and qualities in a soothing scene helped in recovery after a psycho-physiological stress. Moderate depth and complexity, the presence of a focal point, the structural qualities of the scene, and the presence of natural elements such as water and vegetation evoked positive emotions and controlled the occurrence of negative thoughts, which encouraged recovery. According to this theory, human nature would be programmed biologically to respond in a positive way to elements of the environment that signal possibilities of survival and well-being. Aesthetic feelings inspired by natural scenes and their effect on recovery after a stressful event would thus have a genetic basis resulting from the biological evolution of the human species. In this respect, the theory is connected to Edward O. Wilson's concept of biophilia.

The third theory, proposed at the end of 2008 by the psychologist Stephan Mayer and his colleagues, asserts that people can perceive the meaning of their life and its finality through a sense of belonging to the natural world. Social psychology clearly recognizes the fundamental need each person has to belong to a human group, to feel connected to others, and to be recognized as a member of a community in which he or she has a role to play. This statement can easily be extended to nature: individuals have a need, rooted in the biological history of the human species, to be affiliated with and connected to the natural world. This world is of course much larger than an individual universe. When people are inserted in a natural context, this fundamental need to belong is satisfied, thanks to relationships that they experience with the elements of nature. They thus derive a psychological benefit from it. In particular, opportunities to reflect on personal problems and to put them in perspective increase in a natural framework, and this is beneficial to well-being.

APPLICATIONS

This research in environmental psychology has important applications. In the medical realm, for example, hospitals in the United Kingdom are encouraged to create "therapeutic gardens." In fact, they are only rediscovering the principle of the cloisters of medieval monasteries, propitious for meditating while wandering in an island of nature. Cancer patients who frequently have to endure excruciating medical procedures are able to lessen their pain by contemplating images of idyllic natural scenes, such as Victoria Falls in Zambia, and listening to relaxing sounds, such as birds chirping and wind rustling through trees. The exterior and interior architecture of offices accentuates interior gardens, wide vistas, and green plants. Urban planning attempts to maintain green spaces in cities while avoiding urban expansion, which destroys the neighboring rural environment. The new current of green urbanism, for example, promotes gardens on the roofs of buildings, tree-lined parking lots, communal gardens, and even walls of buildings painted green or covered with climbing plants to form walls of vegetation.

This research also enables us to understand what motivates individuals to develop a respectful attitude toward the environment. It has been shown that a positive perception of nature and its psychological benefits promotes the adoption of behaviors that decrease one's ecological footprint of consumption, whether through recycling, buying biodegradable products, using public transportation, actively seeking information on environmental issues, and so on. Furthermore, it is more probable that people who have frequent contact with nature and have had the experience of natural environments in their childhood will adopt eco-responsible behaviors. Attitudes toward ecological issues would then be more connected to the affective bond experienced with nature and to the empathy or emotional affinity one has with it, rather than to a cognitive appreciation of the environment. Stated differently, an experience with nature tends to actualize and to reinforce our innate tendency toward biophilia. It makes us aware that each of us is simply a member, one of many, of a vast natural community. This awareness encourages the adoption of a more respectful attitude toward the environment.

THE EXPLOITATION OF ANIMALS

In the history of humanity, the success of the species has been greatly enabled by domesticated animals, from the hunting or guard dog to the horse, the latter having been used as a means of transportation and as a draft animal. Most families own at least one domestic animal, despite the costs associated with it. People of all ages are fascinated by watching wild animals in nature or in captivity. Zoos continue to attract large numbers of visitors. The English have adopted the expression "horsification of landscapes," since so many of their fields are no longer cultivated and are now used for keeping horses. Hippotherapy, which uses horses to help people develop emotional, physical, and psychomotor skills, has worked wonders. Tales, books, and films for young children very often feature animals that are given human behaviors and characteristics. All cultures, whether traditional or modern, are rich in animal symbolism. Animals are often present in various forms of art and decoration.

Our interactions with the animal world are a very effective means of maintaining a close connection with nature. More dynamic than the vegetational world, animals are more fascinating. Our very widespread tendency to anthropomorphize brings us even closer to them; they become our brothers.

In this chapter, I will analyze several facets of the relationship between modern humans and animals. The relationship is not always without danger, not only in the case of illnesses of animal origin (which we will look at in the following chapter) but also when we consume an excessive amount of animal products. What benefits do humans derive from their interactions with the animal world? As for the animals, do they benefit from the company of humans?

The company of domestic animals such as a dog or a cat has comforting and even therapeutic benefits. We confide in and speak to our domestic animals as if they were human beings and members of the family. In the United States, half the pet dogs receive a Christmas gift each year. Some people show more love toward their dog than to other members of their family. Leona Helmsley, the super-rich woman from New York City, bequeathed a large part of her fortune (the amazing sum of $12 million) to her dog, and specified in her will that it be buried in a grave next to hers. Many people carry a photo of their dog or cat in their wallet along with those of their children.

Recently, several scientific studies have shown that the presence of domestic animals has a positive effect on one's health. Dr. Howard Frumkin, one of the directors of the Centers for Disease Control and Prevention, has compiled astonishing results: in a study of 6,000 Australian patients suffering from cardiovascular disease, those who owned a pet had significantly lower blood pressure and cholesterol and triglyceride levels than did other patients; in another study of several hundred people who had had a heart attack, those who owned a dog were six times more likely to have survived after one year than people without a dog, and this regardless of any physiological differences. In a third study, conducted in England, people who had just acquired a domestic animal, mostly a dog or a cat, clearly had fewer minor health problems in the month after the animal arrived than other people selected at random. This difference continued for ten months only for those who had acquired a dog. The results of this study were confirmed in the United States: people who own a domestic animal, especially if it is a dog, visit their doctor much less frequently. Pets do present risks, however: dog bites are a frequent cause of wounds, sometimes very serious, and cats can transmit illnesses to people, including the cat scratch disease, a bacterial infection, and toxoplasmosis, which is particularly dangerous for pregnant women (this illness is also transmitted by undercooked beef or lamb).

The statistical studies cited above were conceived so as to minimize the effect of factors such as the subjects' eating habits, their social status, and whether they practiced a physical activity, notably walking their pet. These studies show surprising associations between health indicators and the presence of domestic animals, but these associations do not necessarily imply causal relationships, so long as an explanatory mechanism has not been uncovered.

Other experiments have revealed the role of animals in reducing stress. A sampling of women was exposed to a stressful stimulus when they were alone, in the presence of a male or female friend, or in the presence of their dog. Objective measurements of stress levels, such as heartbeat, were then taken. The women accompanied by their dog were by far the least anxious, whereas those who were accompanied by a friend had the highest stress level. In another experiment, simply watching an aquarium for a half hour before dental surgery significantly increased patients' degree of relaxation compared with that of other people who had been asked to remain calmly seated or to look at a poster of a waterfall in a forest for the same period. These experiments were reproduced several times in Europe and in North America, always with the same results.

An explanation of these interesting results introduces two properties specific to many natural situations involving animals. On the one hand, these scenes are, like the river of Greek philosopher Heraclitus in which one never swims twice, always changing while always remaining the same. On the other hand, scenes such as an aquarium, a sleeping dog, or a group of horses grazing in a field elicit a psychological association with a feeling of safety, comfort, and the absence of danger. This tranquility inspires a state of relaxation that favors creative thinking and enables stress levels to be controlled.

An increasing number of retirement homes allow their residents to have pets. The use of animals in therapies to treat psychiatric conditions is beginning to be widespread. Hundreds of medical reports describe a positive social response on the part of depressed or antisocial people when they encounter an animal. Many autistic patients have shown rapid progress when a dog is introduced into their therapy. Medical reports describe almost miraculous scenes during which autistic children respond to the presence of dolphins in captivity. Among these patients, the animal elicits more focused attention, an increase in social interactivity, positive emotions, the nonverbal expression of those emotions, and sometimes even the use of speech. To be able to interact directly with an animal and to touch it greatly increases these effects, whereas simply looking at a picture of an animal or watching an animal from a distance has a lesser impact.

The presence of an animal also increases social attraction: a person with a pet is more frequently approached by strangers than when that person is alone. More generally, promoting affective attachment toward domestic animals is part of a very beneficial public health policy, along the same lines as improving food or encouraging physical activity. The advantages in terms of health and well-being are significant and can function as preventive medicine, which often costs less than curative treatments. And yet, modern and urban life, the mechanization of means of transportation, the destruction of natural ecosystems, and the hectic pace of modern life have pushed animals into the periphery of our lives. This is at least the case for live animals.

Despite all the positive ways a relationship with animals contributes to our well-being, our most frequent contact with them is via our plates. Global meat production increased fivefold between 1950 and 2006, representing a rate of increase twice as high as that of the world's population. Since the beginning of the 2000s, meat consumption has increased much more rapidly in emerging economies than in Europe and North America. China has gone from 20 kilograms (44.4 pounds) of meat per inhabitant per year in 1980 to 52 kilograms (115.5 pounds) in 2007—which nevertheless remains lower than the amounts consumed in Western countries. In the years to come, meat consumption will no doubt continue to increase by at least 2% per year, and will double by 2050. In developing countries, eating meat has become a sign of wealth and prosperity. Half the pork produced in the world is consumed in China, and Brazil is the second-highest consumer of beef, following the United States.

The global population of poultry was 17 billion in 2002. It multiplied by a factor of 15 during the twentieth century, whereas over the same period, the human population "only" quadrupled. The pig population multiplied tenfold during the twentieth century, reaching more than 940 million in 2002. In intensive farming, poultry and pigs are fed exclusively grain and concentrated nutrients, which has enabled such growth. The bovine population, whose food also includes fodder, has increased at the same rate as the human population during the twentieth century, reaching 1.3 billion in 2002. An American or a European on average eats 224 grams (8 ounces) of meat per day—or double the world average, and much more than their minimum nutritional needs—whereas a Chinese person consumes an average of 142 grams (5 ounces). Each year, in a rich country, a family of four eats on average 120 chickens, 4 pigs, and a cow, which represents two-thirds of the protein absorbed by the members of that family.

EATING LESS MEAT

For the inhabitants of rich countries, few behavioral changes would have as many benefits as the adoption of a diet containing less meat and more vegetables. Eating less meat would have many positive effects on the environment, ease malnutrition in poor countries, improve health in rich countries, and be a positive change for animal well-being. The environmental impacts of meat production are enormous and particularly severe in the case of intensive animal farming. In the last few years, the industrial production of animals has increased twice as rapidly as production in traditional animal farming.

Animal farming requires a great amount of space. Pastureland takes up no less than one-quarter of the land not covered by ice. By comparison, cultivated land occupies only 12% of that land surface. The expansion of pastureland in response to the growing demand for meat is often done at the expense of tropical forests rich in biodiversity and which hold an enormous quantity of carbon in the vegetation and the ground. In Latin America, the expansion of large ranches is the principal cause of deforestation, particularly in Brazil, Bolivia, and Paraguay. Furthermore, one-third of the cultivated fields in the world are devoted to raising crops that are used to feed the animals in intensive farming. Here, too, the differences among regions in the world are enormous. Whereas in rich countries 60% of the grain produced is used to feed livestock, populations of poor countries consume their grain directly. Indeed, it would be inconceivable, in places where children are dying of hunger, to feed a pig with that valuable commodity. As for the emerging economies, they are rapidly adopting the Western mode of consumption. China imports increasing quantities of soybeans from Brazil to feed its chickens and pigs. And the expansion of large soybean farming is another important cause of deforestation in the Amazon basin.

Meat consumption is an intrinsically inefficient process, because to obtain 1 calorie of beef from intensive farming, it takes 8–15 calories of vegetable food, which could have been consumed directly by humans. In weight rather than calories, it takes 7 kilograms (15.5 pounds) of grain to produce 1 kilogram (0.4 pounds) of beef. For 1 kilogram of pork, 4 kilograms (8.8 pounds) of grain are necessary. For 1 kilogram of poultry, no more than 2 kilograms (4.4 pounds) of grain are required. The diversion of farmland for meat production is thus enormous, while more than 925 million people in the world suffered from hunger in 2010. This competition between humans and animals for the consumption of grain translates to an increase in the price of the latter, which has tragic consequences for the poorest populations.

A diet rich in meat requires the availability of two to four times more farmland than a vegetarian diet when that meat has been produced in intensive farming, where the animals feed on grain. Intensive livestock raising consumes around 2.5 billion hectares (6.2 billion acres) throughout the world. Consequently, unused productive land that is suitable for farming is becoming rare, notably with the increase in biofuel production and the loss of farmland due to land degradation and the expansion of cities. Since 2008, countries such as China, the Persian Gulf states, and Japan have been encouraging their businesses to acquire the best agricultural land in Africa, South America, and the former Soviet republics of Ukraine, Georgia, Kazakhstan, and Uzbekistan—the three regions of the world that still contain a surplus of land—in anticipation of the moment when all the world's cultivatable land will have been farmed. Humanity could reach that fateful moment around 2050 unless we tolerate a massive tropical deforestation, whose ecological cost would be very high, or if revolutionary biotechnologies are discovered.

Intensive animal farming also has a great impact on water resources: large quantities of water are consumed to irrigate the crop fields and, to a lesser degree, to water the animals. To produce 1 kilogram (0.4 pounds) of beef in intensive farming can require 10,000 to 25,000 liters (2,631.5 to 6,578.9 gallons) of water per year, depending on climate conditions. Furthermore, water pollution near industrial animal farms can be considerable due to the runoff of nitrogen, phosphates, antibiotics, hormones, arsenic, and pesticides into the groundwater. Each day, a pig produces four times more excrement than a human being. A large farm of 5,000 pigs thus produces as much fecal matter as a small town of 20,000 inhabitants. The difference is that most of these farms are not equipped with sewer and water purification systems. The slurry thus produced is spread over the fields, often in amounts that exceed the needs of the crops. All this leads to pollution of surface waters, the groundwater, and the air.

By itself, global meat production contributes to about 18% of the greenhouse gas emissions responsible for climate change, according to estimates that still need to be refined. This figure includes the gases emitted at each stage of the meat production cycle: deforestation to create pastureland, the production and transport of fertilizer, fuel for agricultural machines, and gases from the digestive system of cattle and sheep. As a whole, this sector contributes more to global warming than the entire transportation sector, which represents 13% of greenhouse gas emissions. Animal farming is the main source of methane emissions (CH_4), which result from microbial fermentation in the digestive system of ruminants (the flatus and belches of cows and sheep). Nitrogen protoxide (N_2O), another greenhouse gas, results from the use of fertilizer. These two gases are emitted in lower amounts than carbon dioxide (CO_2), but they have a warming power that is respectively 23 and 296 times higher than CO_2. The amount of time these gases remain in the atmosphere is a dozen years for methane and a hundred years for nitrogen protoxide.

The production of 1 kilogram (0.4 pounds) of beef emits seven times more greenhouse gases than the production of 1 kilogram of chicken, and fifty times more than that of 1 kilogram of vegetables. The production of 1 kilogram of beef in intensive farming has the same impact on the climate as an automobile trip from New York to Boston. When a person adopts a vegetarian diet rather than one rich in meat, the benefits thus engendered for the climate after one year are the same as if that person had replaced his gas car with a hybrid car for the same period. For wealthy populations concerned with reducing their ecological footprint, changing their diet thus deserves as much attention as do modes of transport or the energy consumed by their households. And yet, whereas many industrialized countries subsidize the purchase of hybrid cars and the installation of solar panels, they continue to massively subsidize food crops for livestock (corn and soybeans) and the industrialization of meat production. The beef production sector, which is by far the most polluting in agriculture, is also the one that receives the most subsidies.

The environmental impact of meat consumption does not stop with the agricultural land, the tropical forests, and the atmosphere: as of recently, it also extends to the depths of the oceans. Have you ever sensed that your piece of chicken had a fishy aftertaste? Before being transformed into a piece of meat on your plate, some chickens have been fed, in part, . . . fish. Note, however, the progress here compared to the concentrated food for cows that is enriched in proteins obtained from the carcasses of other cows. The very small fish taken in mass amounts from the oceans by gigantic nets and loaded onto fishing boat-factories are notably transformed into food for livestock, after being pressed to extract their oil. These are primarily sardines, anchovies, and menhadens. These small species represent 37% of the total mass of fish taken from the oceans. Close to half the 30 million metric tons (33 million tons) of small fish caught each year is used to feed the larger fish raised on fish farms; the other half is used as food for pigs (24%) and poultry (22%) in intensive farming, or added to cat and dog food (10%). Have you always thought that Japan, the land of sushi, is the greatest consumer of fish? The global population of pigs and chickens each year consumes twice as much fish as all of Japan. The big problem is that these little fish normally serve as food for marine mammals and birds, as well as for large carnivorous fish. By harvesting the basis of the food chain from the oceans to feed our domestic animals on land, we are condemning the large fish a second time: those that escape the nets of global fishing are seeing their food disappear. Adopting the principle that land animals should be fed from food produced on terra firma would relieve some of the pressure on the oceans.

As in the case of grain, transforming fish into food for pigs and chickens is fundamentally not very efficient: it takes 3 kilograms (6.6 pounds) of small fish to produce 1 kilogram of meat. It would thus be much more efficient to eat these little fish directly; they are, moreover, an important part of the traditional diet of coastal societies. To deprive oneself of a plate of grilled sardines or an anchovy salad in order to be able to consume an industrial chicken with white and tasteless flesh is hardly a boon to gastronomy. As for farm-raised fish, they feed in part on industrial food made out of soybeans, which are notably raised in the Amazon. It is hardly surprising that they taste a bit like tofu! Eating meat empties the oceans, and eating fish accelerates Amazonian deforestation. This is one of the ironic oddities of modern food systems.

Today, average global meat consumption is 100 grams (3.5 ounces) per person per day, but with enormous variations between rich and poor countries. In the poorest countries, ten times less meat is consumed per person than in the richest countries. An amount that would have an acceptable impact in terms of one's ecological footprint and which would be equitable on a global scale would be 90 grams (3.2 ounces) per person per day, on the condition that the meat of ruminants (cows, sheep, and goats) does not go beyond 50 grams (1.7 ounces), the rest coming from monogastric animals (chickens and pigs), which emit less methane. The current average in the richest countries, let's recall, is 224 grams (8 ounces) of meat per person per day, which shows the extent of the effort to be undertaken: we should eat 2.5 times less meat than we are eating today. Such a decrease would enable populations of the poorest countries to increase their consumption to a reasonable level while stabilizing the meat production sector's contribution to greenhouse gas emissions. Furthermore, it would have a positive impact on health in the rich countries, where an excessive amount of meat is consumed.

HEALTHY EATING

Meat is not an indispensable component of a human being's diet. Nevertheless, it provides a series of important nutritional elements that are not easily obtained through vegetables, in particular iron, zinc, and lysine. However, the protein contained in meat is not of higher quality than that in vegetables. Meat is the main source of saturated fats, which provide no health benefits. Consuming a lot of that fat has been associated with an increase in the cholesterol level in the blood.

Even in developing countries, food is becoming increasingly rich in fat, which today represents 25 to 30% of ingested calories. The traditional diet of rural areas, based on vegetables and with very little meat and an absence of sugar (with the exception of fruit and honey), is replaced in cities by food that is rich in animal fat and contains less fiber and a lot of sugar. Throughout the world, vegetarians suffer less from obesity and excess weight than meat eaters. This is only a correlation, as factors other than diet can be the cause of this, because the fact of being a vegetarian is associated with other individual and contextual characteristics that can also have an effect on health. A 2010 study of almost

400,000 European adults found that eating meat was linked with weight gain, even in people taking in the same number of calories and having a similar level of physical activity. An extra 250 grams (8.9 ounces) of meat a day—equal to a small steak—led to an additional weight gain of 2 kilograms (4.4 pounds) over five years.

A diet rich in meat has been plausibly associated—though this is not certain in the current stage of the research—with an increased risk of colon cancer (or colorectal cancer), which is most widespread, and, more hypothetically, with other cancers. According to some studies, the risk of colorectal cancer decreases by around a third each time a daily consumption of red meat is reduced by 100 grams (3.5 ounces). The consumption of processed meats (bacon, sausages, hot dogs, salami, ham and smoked meat, or any cured meat) has been associated with an increase in the risk of stomach cancer, perhaps due to the nitrites that are added to prolong their preservation. It would, however, be premature to conclude that there is a relationship of cause and effect. Animals permanently raised in pens accumulate fatty omega-6 acids (the "bad" fats), associated with cancer, diabetes, and obesity. Countries where large quantities of meat are eaten, such as Argentina and Uruguay, where you are served grilled meat for breakfast, are also countries where the rate of colon cancer is the highest in the world. Furthermore, the risk of coronary or cardiac diseases probably increases with an excessive consumption of meat or, in any case, of the fats that it contains.

Meat increases the risk of ingesting traces of antibiotics, which are often added to the concentrated food given to cows and chickens in industrial farming. Indeed, in these farming conditions the constant closeness of the animals increases the risk of infection and contagion among them. They are selected based on criteria of productivity, which inevitably leads to a decrease in genetic diversity within the population. The immune systems of all the animals in the same farm thus respond in the same way to a pathogen: a good mutation of a virus or a bacterium is likely to kill all the animals in a few days. In this context, only the systematic use of antibiotics enables such epidemics to be avoided. From 25 to 75% of those administered to animals are found in the fertilizer and thus in the environment via the soil and water. This favors the appearance of resistant bacteria, which threatens the efficacy of some treatments of human illnesses—a phenomenon that contributes to resistances that occur due to an overuse of antibiotics prescribed by doctors. In the United States and in some European countries, around one-third of the pigs are carriers of a new strain of methicillin-resistant golden staphylococcus (MRSA), a bacterium against which the most common antibiotics have lost their efficacy, and which is responsible for nosocomial illnesses (illnesses contracted in hospitals).

Persistent organic pollutants, that is, pesticides and toxic products of industrial origin such as dioxins, furans, and polychlorobiphenyls (PCBs), accumulate in the fatty tissue of animals and enter into the human diet through them. Concentrations of these toxic molecules become very high when the animal fat recycled through the carcasses of slaughtered animals enters into the production of livestock food. These persistent organic pollutants are also found in the flesh of farm-raised fish, which are fed concentrated food notably made from animal protein. These molecules are carcinogenic and toxic to the development of the nervous system in a fetus or young child.

The variant of Creutzfeldt-Jakob disease linked to a bovine prion—a human variant of the bovine spongiform encephalopathy (BSE), more commonly known as mad cow disease—was identified in Great Britain in 1996, ten years after the first appearances of BSE cases in that country. Since then it has been detected in several countries, including France. The prions responsible for this illness seem to have been transmitted to humans through infected beef. The consumption of meat also increases the risk of exposure to bacteria such as salmonella or *Escherichia coli* O157:H7, a dangerous strain of which is found in a natural state in the intestines of livestock, poultry, and other animals (whereas many other strains of this bacterium pose no threat). Several zoonoses, which we will discuss in the following chapter, are also associated with intensive animal farming.

THE GIANT HECATOMB

In ancient Greece, during the annual feast of the Panathenaea, cities participated in the religious sacrifice of one hundred cattle, called a hecatomb. Today, more than 56 billion domesticated animals throughout the world are killed each year so that humans can consume their meat. For many of them, going through the slaughterhouse means deliverance from a life that was rarely a long, peaceful road. Animal farming is increasingly concentrated in huge industrial animal farms. This is the case for 43% of cattle and more than 50% of pigs and chickens. In the United States, half the production of poultry is controlled by only five companies. In these true factories, animals endure conditions of captivity, immobility, and overpopulation. For their entire life they have little contact with natural light, never touch the ground of a field, and ingest only concentrated food enriched in protein. In addition to a lack of living space, these animals breathe a pestilential air filled with ammonia, dust, and microorganisms, their feet permanently standing in their excrement. The harshest living conditions are imposed on laying hens, transformed into exploited machines until they die.

Transport to the slaughterhouse can also be the source of a great deal of stress for the animals. Due to the frenzied rhythm of these conveyor-belt factories, the killing operation at slaughterhouses experiences occasional failures. Although they aren't common, these situations can cause the animals extreme suffering. In the case of an epizootic disease, like the avian flu, the birds are killed on a large scale by inexperienced people without the appropriate equipment, in conditions that are not always in conformity with established guidelines, in order to rapidly control the propagation of the illness.

Most meat production today occurs in developing countries, where legislation on animal well-being is lax, a culture that respects animals is often absent in the meat industry, and the funds allotted for inspections insufficient. In these countries, pressure from the civil society to abolish the bad treatment inflicted on animals is almost nonexistent.

Can humans construct their well-being on the suffering of other living beings? Is it possible to be happy when, on a large scale, animals endure unbearable living conditions? Illustrious men have already responded to this question. In the seventeenth century, Descartes asserted that animals had no soul or mind, and could thus be considered simple machines. According to him, "since beasts do not speak as we do," they do not have thoughts.[1] By contrast, in the third century BC the Greek philosopher Theophrastus, disagreeing with his master Aristotle, was against the consumption of meat, because it deprived the animals of their life, and was thus unjust. He thought that animals could reason, feel, and sense just like human beings. St. Francis of Assisi, known for his compassion toward animals, asked "all brothers in the world that they respect, that they venerate, and that they honor all living things. Rather, all that exists."[2] The eighteenth-century English philosopher Jeremy Bentham wrote that the suffering of animals was as real and morally important as human suffering. Shortly after France had granted fundamental rights to its former slaves, he declared, "The day may come, when the rest of the animal creation may acquire those rights which never could have been withholden from them but by the hand of tyranny."[3] Tolstoy preached a respect for life in all its forms and became a vegetarian. He held the conviction that by killing animals, "man suppresses in himself, unnecessarily, the highest spiritual capacity, that of sympa-

1. From R. Descartes, *Discours de la Méthode pour bien conduire sa raison et chercher la vérité dans les sciences*, part 5.

2. French quote found at http://www.buddhaline.net/spip.php?article756.

3. J. Bentham, *An Introduction to the Principles of Morals and Legislation*, found on http://www .utilitarianism.com/jeremybentham.html.—Trans.

thy and pity towards living creatures like himself, and by violating his own feelings becomes cruel."[4] Eating animal flesh was, according to Tolstoy, absolutely immoral. Gandhi is thought to have declared, "The greatness of a nation and its moral progress can be judged by the way its animals are treated."[5] Albert Einstein is counted among other illustrious vegetarians. Marguerite Yourcenar had the hero of *L'Oeuvre au noir* say, "I refuse to ingest agony." In his novel *Elisabeth Costello*, Nobel Laureate in Literature John Maxwell Coetzee, another vegetarian, criticizes the massive exploitation of animals that is carried out in complete indifference. His heroine readily compares current society, in which the daily massacre of animals is common, to the situation under the Third Reich, when people living near the concentration camps knew or at least suspected what was going on. In Buddhism, animals are considered sentient beings that must be respected and loved, as is seen in many sutras.

ANIMAL RIGHTS

The recent evolution of value systems in Western societies has rejected racism, sexism, and homophobia—even if these forms of discrimination persist in many places. Isn't the next great moral challenge to reject "speciesism," that is, all forms of discrimination that involve species? Speciesism is to the species what racism and sexism are respectively to race and gender: the will to ignore (or to ignore more) the interests of some to the benefit of others. Anti-speciesism is against the exploitation and bad treatment by humans of individuals of other species. Today it takes the form of multiple movements for the defense of the well-being and rights of animals, and even for their "liberation."

Following Jeremy Bentham, Peter Singer, a philosopher at Princeton University, proposed the following moral argument to justify the liberation of animals: in the Western world, the dominant ethics considers that the interests of humans should always win out over the comparable interests of members of other species; however, despite the incontestable differences between humans and nonhuman animals, we have in common the ability to suffer and to be happy—which means that like humans, animals have their own interests, notably that of not suffering, a common point that justifies equal treatment with regard to suffering. The moral claims benefiting animals are thus founded

4. From L. Tolstoy, *The First Step*, in *Essays and Letters* (New York: H. Frowde, 1909), 82–91.

5. This quote is widely attributed to Gandhi, sometimes citing Ramachandra Krishna Prabhu, *The Moral Basis of Vegetarianism* (1959), but this has not been verified (R. Keyes, *The Quote Verifier* [New York: St. Martin's Press, 2006], 74).

on the interests that they share with the human species and not on so-called equivalent abilities.

To think that humans are superior to animals because they have the ability to reason and are aware of their own existence is hardly defensible, because infants or the profoundly disabled—the "marginal cases," in the language of moral philosophy—have these attributes to a lesser degree than many animal species. According to Peter Singer, if we draw a circle inside which we place beings whose lives deserve respect and exclude those who can be exploited to our own advantage, and if we place the people of our species suffering from a profound cognitive disability inside that circle, then all the sentient animal species capable of suffering must deserve an equivalent degree of respect. This certainly includes mammals, birds, fish, reptiles, amphibians, and some of the shellfishes (the crustaceans). Insects, spiders, worms, and most mollusks, as well as more elementary forms of life, are probably outside that circle—eradicating mosquitoes and ticks that spread illnesses such as malaria and Lyme disease would thus be morally acceptable. The species of this latter group, which could morally be eaten, are, however, rarely part of the human diet, with the exception of some mollusks (oysters, mussels, clams, and scallops). To place the animals belonging to the first group outside the circle of beings whose lives deserve respect simply because they are not part of our species is as discriminatory as racism and sexism; that would result, according to Singer, in committing the same moral sin as the White or the man who considers himself morally superior to the Black or to a woman simply because of his race or gender, regardless of the other's intrinsic qualities—speciesism is the same thing.

The human genome and that of chimpanzees are almost identical: they differ by only 1.2%. The results of recent biological discoveries now attribute to animal species, particularly primates, characteristics that we thought only a short time ago to be the privilege of humans: culture, the use of tools, a form of language, complex social organizations, and even, perhaps, self-awareness. Research on the cognitive capabilities of animals is still in the very early stages. The true justification for an exploitation of animals is that it serves our interests as the dominant group, in the same way that slavery and sexism served the interests of our ancestors. Aristotle wrote that barbarians were human beings who existed to serve the good of the more rational Greeks. Slave owners considered Africans to be human beings because they had converted them to Christianity, but that did not prevent them from treating them as mere merchandise. The dominant group adapts its moral views concerning the beings to which it grants no rights and which it exploits without scruples in function of its own needs, whether or not they belong to the same race or the same species: in the past, the barbarian or African slaves; today, animals. An attempt to justify speciesism is only a way of covering an immoral act with an argument that is difficult to defend.

The notion of extending moral concerns to animals inspired violent criticism against Peter Singer's thesis, which has not always been well understood and which, it is true, is fairly controversial. He was, for example, accused of defending Nazi theories, because blurring the border between humans and animals could justify the inhuman treatments inflicted on the weakest humans. Singer stressed, of course, that it is not a matter of lowering the status of humans but of raising that of animals. Other philosophers have suggested that the widespread conviction that there exist fundamental human rights necessarily leads to an acceptance that similar rights should be applied to animals. Some conservative Christian figures—including Matthew Scully, a former speechwriter for George W. Bush and the author of a noted book on cruelty to animals entitled *Dominion: The Power of Men, the Suffering of Animals, and the Call to Mercy*—preach a respect and a dignity for animals as living beings created by God.

Let's note that these arguments support the view that inflicting needless suffering on animals and treating them as simple units of production in a capitalist system is not morally acceptable. This doesn't necessary rule out the consumption of meat, on the condition that the animals are provided with good living conditions and that the slaughter be carried out in settings that do not inflict physical or mental suffering. Here is where the difference between the defense of animal well-being, which aims to protect animals from suffering and ill treatment, and the promotion of animal rights, which can go as far as an opposition to any form of "use," arises. The case of laboratory animals enters into an entirely different category when the experiments to which they are subjected are likely to lead to treatments that will save many human lives. Any form of needless suffering must also be banned here. Animal experimentation to test the danger of cosmetic products, for example, cannot be defended, given the superfluous nature of those products. In short, an intermediate position must be found between speciesism, which denies any rights to nonhuman species, and the most extreme positions, such as "biocentrism," which thinks that all organisms on earth are equal and should receive the same degree of consideration and the same moral rights. This last belief inspires some terrorist movements for the liberation of animals and the earth.

ADVANCES IN ANIMAL WELL-BEING

Since the end of the 1970s, significant progress has been made in the struggle against the suffering and ill treatment of animals. The European Commission has played a pioneering role in this realm, aiming for a respect of five protections: from discomfort; hunger and thirst; fear and distress; pain, suffering, and illnesses; and the impossibility of expressing a natural behavior. The European Union has put many practical measures in place to assure real progress in this realm, through directives that are then translated into national legislation. The number of laboratory animals used in research for experiments has decreased by around half since the 1960s. Since 2007, European directives have regulated conditions for raising baby cows, notably prohibiting battery farming. By 2012, egg producers in the European Union will have to increase the minimum size of the laying hens' cages to at least 750 square centimeters (115.3 square inches) per animal, which will represent an increase of 60% in most cases—cages today are so small that the chickens are unable to beat their wings or turn around. In 2004 in Paris, the World Organization for Animal Health, which in some ways is the animal equivalent of the World Health Organization, held the first world conference on animal well-being, and the following year adopted global principles in this area.

In 2002, Germany was the first country in the European Union to include animal rights in its constitution. At the Bundestag, legislators approved by a two-thirds majority that the words "and animals" be added to the constitutional clause requiring the state "to respect and protect the dignity of humans." In June 2008, Spain initiated a legislative procedure for the defense of large apes through a resolution that granted them rights similar to those of humans. In addition, Spain convinced the European Commission to extend that measure to all twenty-seven member countries. In 2009, Ecuador moved one step further by including in its constitution the granting to nature—tropical forests, rivers, air, and the diverse fauna that depend on it—the same inalienable rights to exist and flourish as those enjoyed by human beings.

The response of certain laboratories in the agro-food industry has been less promising: for example, research is aiming to create a genetically modified pig in which the genes responsible for stress response would be neutralized. The idea is thus to produce a new pig able to endure horrible conditions of industrial farming and slaughtering without feeling any fear, which would not, however, prevent it from experiencing physical pain. This research thus aims to suppress any instinct for survival in the genetic material of pigs. The motivation of the researchers is to ensure the quality of the meat, not to diminish the suffering of the animal. Indeed, a pig's reaction to stress during transport and slaughtering influences the speed of the muscular metabolism before and after being put to death, and thus affects meat quality. And so the animals are allowed to recover for several hours between their transport and their slaughter in order to obtain quality meat, which then increases prices. As Matthew Scully would point out, rather than rethinking a mode of industrial farming adapted to the animal, researchers are creating a pig adapted to industrial farming.

The progress made in the realm of animal well-being in Europe and North America has had an unfortunately minor impact, because most meat production today occurs in developing countries, and this will increasingly be the case in the future. Furthermore, the rules of the World Trade Organization (WTO) work against this progress by prohibiting a refusal to import animal products from a country that does not respect the international standards of animal well-being. Indeed, the WTO does not accept modes of production as a criterion for trade discrimination.

Surprisingly, the cause of animal well-being does not find a lot of support among the public. The agro-food industry tries to hide the reality of farming conditions, of slaughtering, and of meat preparation. Without that concealment, the number of vegetarians would no doubt explode overnight. The day when a majority of consumers will buy only meat and other animal products that have come from sources certified for promoting a respect for animal well-being, which no doubt would imply an increase in prices, a significant improvement will be seen. Private enterprises downstream from the meat production chain also have a role to play, particularly when they operate throughout the world. Recently, McDonald's has started buying its meat and its eggs only from sources that exclude cruelty, abuse, and negligence in the farming, transport, and slaughtering of the animals. Norms of animal well-being have been defined by a council of independent experts on the protection of animals. Companies all along the meat production chain that provide to McDonald's are the objects of audits by inspectors, also independent. Direct competitors have rapidly made similar changes. Vegetarian burgers have also been introduced into some countries.

OUR LEISURE, NOT THAT OF ANIMALS

Since recent research in medicine and psychology has shown the beneficial impact of pets on human health, couldn't we promote other ways of interacting with the animal world than by ingesting their meat? Animals are an integral part of some of our leisure activities, from hunting to horseback riding; they participate in entertainment spectacles, such as bullfighting and circuses and rodeos. If by definition these leisure-time activities make us happier, do they provide an equivalent benefit to the animals who participate in them?

Bullfighting is difficult to defend, because only a desire to perpetuate a local tradition can justify this spectacle, which leads to the suffering of an animal and displays the all-powerfulness of man confronted with a brutal beast. Regarding circuses, several northern European countries have voted on legislation forbidding or regulating the use of wild animals. In the United States, zoos, circuses, and marine mammal parks are regulated under the Animal Welfare Act, which requires that minimum standards of care and treatment be provided to animals. Hunting is sometimes a true necessity, and is then justified, whether in the name of a need for food in poor countries, or the need to control game populations in wealthy countries. It is very common, all the same, for the game to be raised and fed by people, first in a cage, then in a natural environment, before being hunted. Each year France farms some tens of millions of eggs of game birds, most destined to be hunted. From the point of view of animal well-being, those conditions are the same, or better, than those that prevail in industrial farms and slaughterhouses. From the human point of view, in a leisure activity practiced in that way, there is only the pleasure of killing. To enjoy nature with friends does not mean you have to have a rifle on your shoulder. Target shooting can be practiced just as well on nonliving targets. Great Britain in 2004 was a pioneer in legislation for the well-being of animals by prohibiting hunting with hounds, a particularly cruel practice.

As for the quest for trophies during safaris in South Africa, such hunts are very far from a natural experience: the "wild" animals are raised on confined ranches that resemble zoos. This is in fact called game farming. All species of wild animals, from rhinoceroses to giraffes, are sold at auction among farmers during a weekly Internet sale and transported throughout the continent after undergoing strict veterinary controls. The price of an animal depends in part on the number of infections it is carrying. Hunters then place their orders from abroad and come to shoot the animals that have been previously identified and then followed daily by a GPS system, and toward which the clients are guided for the fatal shot. The stuffing and shipping of the trophy are part of the cost of the all-inclusive vacation, in which it is, above all, the staging of an adventurous experience that justifies the astronomical sums paid.

WHICH IS HAPPIER, THE HORSE OR THE ZEBRA?

Whereas the zebra is one of the species sometimes hunted and most often photographed during African safaris, the horse is at the center of very popular sporting and leisure activities. Can we say that the zebra, which enjoys a great deal of freedom, is a happier animal than the horse, which is subjugated by humans? Isn't nature domesticated by humans a servile and debased form of wild nature? For a domesticated animal, what does "the possibility of expressing its natural behavior," which is one of the protections of animals promoted by the European Commission, really mean? In La Fontaine's fable "The Wolf and the Dog," the wolf, who is skin and bones, no longer envies the dog's plumpness after he discovers, upon seeing his raw neck, that the dog lives chained up.

Wild horses, zebras, and the domesticated horse are today distinct species, which come from the same branch in the biological world. A domesticated animal is a species selectively raised in captivity, and thus modified compared to its wild ancestors in order to be more useful to humans. The domesticated horse has existed in its current form only for a bit more than five thousand years, which supports the hypothesis that it is the fruit of a selection made by humans—a selection that is pursued, moreover, in a very active way today. The zebra, on the other hand, has never been domesticated and has always been very difficult to tame.

In Africa, colonists made several attempts to tame zebras in order to take advantage of their strong resistance to heat and tropical illnesses. Belgian lieutenant Fernand Nys had been sent to the Congo in 1902 with the mission of studying their training. Out of 90 zebras captured in Katanga, 30 died after a few days, either because they refused to drink from the receptacles touched by humans, or because they caused their own death by throwing themselves against the walls of their stable. The zebra is much more unpredictable than the horse: it panics easily and, as an adult, has the bad habit of biting and not letting go. At the end of the nineteenth century, Lionel Walter Rothschild, from the famous banking family, created a sensation in the streets of London by traveling in a carriage pulled by four, albeit young, zebras.

Of course, those few recent and brief attempts to domesticate zebras cannot be compared to the domestication of the horse, which has occurred over several thousand years. Horses have played an essential role in many civilizations. They have been used for transportation, war, pulling plows, hunting, and sport. Societies that have mastered the horse have benefited from a major competitive advantage—to be convinced of this, one need only think of the vast empire conquered by the Mongols in the thirteenth century and of the conquest of the American continent by the Spanish in the sixteenth century. More than a useful animal, the horse has always been a noble and respected companion.

And in rightful return for the services rendered to humans, has the horse been paid with more well-being? But what is well-being, or happiness, for a horse, or for an animal in general? According to American journalist and writer Michael Pollan, for an animal, happiness consists of having the opportunity to express the nature of its species or, according to Aristotle, of following "the characteristic form of life of each creature": for the lion, it means being able to live as a feline on the African savannahs; for the bird, to be able to fly; for the pig, to live in a piglike way; and for the horse, like a horse. And for all domesticated animals, the very nature of the species is to have a destiny intimately connected to that of humans, as a companion (the dog) or as a source of meat (the pig and the cow).

Domesticated species are the products of a biological evolution directed by human selection. It involves a contract that benefits both parties: humans use the domesticated animal, and the animal is associated opportunistically with humans in order to increase its chances of survival as well as those of its offspring. The domesticated animal species benefits from this on the whole. Granted, humans have subjugated the horse, but the horse has also brought humans into its service. Indeed, humans spend a lot of energy sheltering, providing food and drink for, taking care of, and training the horse, whereas the zebra is left to itself to face the droughts, illnesses, and predators. While the zebra lives in a constant state of alertness against nocturnal attacks by felines, the horse sleeps peacefully in a straw-padded stall.

Today there are more than 60 million horses on the earth, from the Nordic regions where they spend the winter in heated stables, to Dubai where the stables are air-conditioned. On the other hand, of the three species of zebras, two are in danger of extinction and the third, the most common, is seeing its population decrease rapidly due to hunting. The population of the Grevy's zebra, or imperial zebra—the largest of the species—has gone from 15,000 a few decades ago to fewer than 2,000 in the wild today. Confined on the plains of Somalia, Ethiopia, and northern Kenya, it is one of the African mammals whose population has decreased most rapidly.

The greatest disaster for cattle, pigs, and chickens would be if the entire human population becomes vegetarian or, even worse, vegan—the practice that excludes all animal products, including eggs, milk products, honey, and so on: those animal species would then disappear from the face of the earth. During the twentieth century, cattle and pigs multiplied at an unprecedented rate in the animal world. Unfortunately for those species, this enormous demographic success is not dissociated from the industrial methods of farming that, starting in the 1960s, have inflicted the deplorable living conditions already mentioned on the animals destined for human consumption. For them, domestication has been transformed into a trap with the advent of intensive animal farming and slaughtering some tens of thousands of years later. The battle to be fought is thus for the improvement of animal well-being in intensive farming, not for the suppression of domesticated species that would become useless and cumbersome for humans who become vegans. A happy life and merciful death were the two conditions that, for the philosopher Jeremy Bentham, justified the consumption of meat.[6]

Raised more often for sport and leisure than for consumption, does the horse have any reason to regret its alliance with humans? It was known in 1200 BC that the horses of Anatolia lived in buildings that were better maintained than the lodgings of their grooms. The royal or imperial stables of the great courts of Europe, from Versailles to Vienna, were true palaces, today transformed into museums. As far as farms of the past are concerned, the horse was well treated, because it enabled fields to be worked, merchandise to be transported, and people to move around quickly. The contemporary practice of riding does not, however, escape from an honest look at our consciences. The greatest rider in the twentieth century, Portuguese master Nuno Oliveira, told his students, "ride with your heart," "feel and ride with emotion," treat the horse so that it will move willingly, and never use force. He defined the trained horse as a happy horse and riding as a dialogue with it, a quest for understanding and perfection. He asked everyone, after setting their feet back on the ground, to look attentively in the horse's eye to see if he seemed happy.

And yet, the risk of an abuse of horses is never far away, including during competitions at the highest level in various equestrian disciplines. Racing causes fatal accidents for more than 400 horses each year. In 2008 in eventing, the culmination of equestrian competition, a rider with several Olympic medals was seen putting protective leggings on a horse's legs, inside which were tiny nails. The rider, who was not sanctioned due to a lack of sufficient proof, participated in the Olympic Games in Peking a few weeks later. An Olympic champion in jumping, a triple world champion, was disqualified from those same games for doping his horse. The product used was a derivative of hot pepper, which has hypersensitizing properties and provokes a burning sensation in the horse's legs, which makes it raise its legs above the fences. In dressage, the technique of neck hyperflexion, widely practiced by the greatest champions, was forbidden at the end of 2008, because it threatened the horses' well-being. The new regulations of the International Equestrian Federation specify that "the object of dressage is the development of the horse into a happy athlete through harmonious education."[7] Equestrianism cannot survive as a sport and leisure activity unless it maintains a positive image with the public and thus respect for the horse.

6. See J. Bentham, *An Introduction to the Principles of Morals and Legislation* (1780; New York: Dover, 2007).

7. See http://twww.fei.org/sites/default/files/file/DISCIPLINES/DRESSAGE/Rules/RULES _DRESSAGE_2011_BLACK-VERSION_web.pdf.

Whatever the form of the relationship that humans maintain with animals, they derive many benefits from it. Whether the animals are domesticated or wild, used as pets or raised for economic ends ("commercial animals"), a close contact with them has a positive influence on our well-being and health. An interaction with animals is an important component of the ecology of happiness. By contrast, reducing this interaction to an excessive consumption of meat has harmful consequences, not only for the environment, our health, and our ability to feed all of humanity decently and equitably, but also on animal well-being. A more developed moral sense should push modern humans to form less unilateral and more respectful relationships with the animal world, whether wild or domesticated. The future, the survival, and the well-being of animals are entirely in the hands of human societies, as we have acquired almost total control over the fate of wild animals by greatly shrinking their natural habitats; this necessarily means that we must take greater responsibility for them.

Just as in our relationships among fellow humans, our interaction with animals must include an altruistic and not only an egocentric dimension. Granting rights to animals can contribute to human happiness by guaranteeing the protection of the animals we love and a better awareness of their well-being. Too often, a concern for animal well-being is secondary to the demands of profitability and performance. When humanity will have eradicated extreme poverty and faced climate change and the degradation of the natural environment, its next great social cause will be that of the well-being and rights of animals. The happiness of some cannot be built on the unhappiness of others.

3

AN INFECTIOUS ENVIRONMENT

It is normal to consider health our most precious possession, especially once it has been lost. A long time ago, Plato believed health to be the ultimate possession, followed by beauty and then wealth. Good physical and mental health is a condition for happiness. Many surveys of a subjective perception of one's level of satisfaction show that those who have health problems are on average less happy than others. This statistical association between health and happiness remains strong, even when all other factors influencing satisfaction are taken into account: a decline in health leads to negative and lasting effects on happiness, through both the impact ill health has on daily life, and the loss of opportunities that can result from it.

Human interactions with natural environments are associated with health in multiple ways. For example, human use of plants as medicines dates to at least 60,000 years ago, as evidenced by fossil records. According to the World Health Organization (WHO), almost 65% of the world's population has incorporated plants into their primary modality of health care. The screening for biological activity of the approximately 250,000 higher plant species is a major source of drug discovery. Nature conservation is thus essential to maintain this still largely unexplored source of potential cures for diseases.

Human-environment interactions also have much less visible health implications, which are the subject of this chapter and the next. For several decades, environmental changes have been related to the emergence of many new infectious diseases. In addition, animals don't have just positive effects on the health and well-being of humans: whereas on a day-to-day basis their comforting and soothing virtues are undeniable, over the long term they have been at the origin of many human diseases. Do certain forms of environmental change threaten happiness through their effects on our health?

What do the following diseases have in common: Creutzfeldt-Jakob disease, a new variant of which is caused by the agent responsible for mad cow disease; Crimean-Congo hemorrhagic fever, often fatal, transmitted by a tick that is found primarily in central Turkey; Lassa hemorrhagic fever, transmitted by a virus endemic to several countries of West Africa, where it infects from 100,000 to 300,000 people every year, 5,000 to 6,000 of whom die of it; Chikungunya disease, which has endured since March 2005 on Reunion Island and on the east coast of India, and whose name means "that which bends up" in the Makonde language, because of the great pain in the joints that it causes—it made an incursion into southern Europe, in Italy, in the summer of 2007. To that list we might add West Nile virus, which, starting from New York in 1999, has spread throughout the United States in only a few years; Ebola fever, which leads to a horrible death from internal and external hemorrhaging, vomiting, and diarrhea; the *Escherichia coli* O157:H7 bacteria, responsible for several pathologies transmitted to humans through the consumption of contaminated food; Lyme disease, transmitted by a tick, an important locus of which was identified for the first time in the United States in 1975. And let's not forget the H5N1 virus of bird flu, a source of much concern in the public health sector, because it was seen a few years ago as a possible cause of the next great pandemic—until the H1N1 virus of swine flu became the main source of worries; tick encephalitis, a fatal illness that occurs in central and northern Europe; the *Legionella pneumophila* bacteria that is spread through air-conditioning systems and causes Legionnaires' disease, identified for the first time in 1976 among members of the American Legion; the hantavirus, at the origin in Europe of a hemorrhagic fever affecting the kidneys, a fatal variant of which exists in North America; the Guanarito and Sabia viruses, also sources of hemorrhagic fevers in Venezuela and Brazil; the Nipah virus, identified only in 1999; severe acute respiratory syndrome (SARS), which caused an abrupt decrease in Asian economies during the first months of 2003; the simian orthopox virus, also called monkeypox; and many more. They are all new diseases that have appeared or been identified during the last few decades, and they all have very serious consequences for human health, some even causing death.

Since 1970 some forty emerging diseases, defined as previously unknown illnesses that affect humans, have taken the world by surprise, such as acquired immunodeficiency syndrome (AIDS) or Ebola. By contrast, reemerging diseases are illnesses that are already known, but their incidence is either growing or their geographic area is expanding; their impact on public health remains great, even if they are less covered in the media than emerging diseases. In its annual report of 2007, the WHO rang the alarm: the rate of the appearance of new diseases, without precedent in human history, has accelerated between 1940 and the end of the 1980s. This was confirmed in a wide-reaching study published in 2008 in the prestigious journal *Nature*, led by a consortium of English and American biologists. The rapidity with which these diseases are surging has caught the medical world off guard. On average, more than one new illness is discovered every year. For example, during the summer of 2008, a mysterious and unknown illness decimated a tribe of Waraos Indians who live in northeastern Venezuela. The symptoms include a partial paralysis of the body, convulsions, an extreme aversion to water, and death after about a month. Even more recently—in the spring of 2009—a new deadly virus was identified, one that causes bleeding comparable to that of the Ebola virus. It was named "Lujo," because its first victims were infected at Lusaka and in Johannesburg. Very aggressive, it can be transmitted by rodents.

These emerging infectious diseases threaten the world economic system and international security. In 1996, the then US president Bill Clinton spoke of them as being "one of the most significant health and security challenges facing the global community."[1] Even the American Central Intelligence Agency (CIA) since 2000 has been worried about the security danger they represent, through their ability to cause not only a high death rate in the United States but also economic decline, social fragmentation, and political destabilization in regions in the world that are most affected, such as sub-Saharan Africa. The emergence of these diseases is linked to several factors, the most significant of which are changes in the natural environment caused by human activity. Since these ecological factors are not among the usual concerns of public health institutions, they have rarely been the object of special studies.

However, the idea that conditions in the natural environment have an important impact on human health is not new. The founder of medicine, Hippocrates, in the fourth century BC introduced his famous *Treatise on Airs, Waters, Places* as follows:

Whoever would study medicine aright must learn of the following subjects. First he must consider the effect of each of the seasons of the year and the

---

1. The White House, Office of Science and Technology Policy, "Fact Sheet: Addressing the Threat of Emerging Infectious Diseases," June 12, 1996; online at www.fas.org/irp/offdocs/pdd_ntsc7.htm.

differences between them. Secondly he must study the warm and the cold winds, both those which are common to every country and those peculiar to a particular locality. Lastly, the effect of water on the health must not be forgotten. Just as it varies in taste and when weighed, so does its effect on the body vary as well. When, therefore, a physician comes to a district previously unknown to him, he should consider both its situation and its aspect to the winds. The effect of any town upon the health of its population varies accordingly as it faces north or south, east or west. This is of the greatest importance. Similarly, the nature of the water supply must be considered; is it marshy and soft, hard as it is when it flows from high and rocky ground, or salty with a hardness which is permanent? Then think of the soil, whether it be bare and waterless or thickly covered with vegetation and well-watered; whether in a hollow and stifling, or exposed and cold. Lastly consider the life of the inhabitants themselves; are they heavy drinkers and eaters and consequently unable to stand fatigue, or, being fond of work and exercise, eat wisely but drink sparely? Each of these subjects must be studied.[2]

The recent environmental changes that have facilitated the emergence of new diseases include climate change, deforestation, the intensification of farming, and, more generally, a modification of the way in which humans interact with the natural environment, either through a colonization of natural spaces that had been previously untouched, or through increased contact with wild or domesticated animals. Of course, environmental changes are not the only causes of the emergence of diseases: socioeconomic changes such as a modification in human demography and lifestyles, the increased frequency of long-distance travel, the development of international trade, and new methods of industrial production made possible through technological progress have also played an important role. The genetic adaptation of microbes and their increasing resistance to medicine are important biological factors. The ineffectiveness or even the collapse of public health policies in some countries, in particular in sub-Saharan Africa, has also facilitated the emergence or the reemergence of diseases. Finally, and paradoxically, by prolonging life expectancy the successes of modern medicine have increased the population of elderly people, whose immune systems are weaker. This has caused a considerable increase in the num-

2. From *Hippocratic Writings* by Hippocrates and J. Chadwick, trans. J. Chadwick and W. N. Mann (Penguin, 1983 [online preview; no page numbers given. See http://books.google.com/books?id=r7EoQzZn7OcC&printsec=frontcover&vq=seasons#v=onepage&q=seasons&f=false]).

ber of people very likely to be infected by new pathogenic agents. Thus, there is a complex interaction between old age, health, illness, and death.

In a 2006 report, the WHO estimated that environmental factors are responsible for at least 24% of the burden of diseases on a global scale. That burden is measured by the total number of years lived in good health lost for all the world's population. Of course, links of cause and effect between the environment and health are not always direct. Environmental changes increase the risk for the appearance of 85 of the 102 major diseases recorded by the WHO, among which "old" illnesses, such as diarrhea, respiratory infections, and malaria, occur most often. The impact of this affects above all populations in developing countries, and in particular children.

It is reasonable to wonder whether infectious diseases represent a mechanism of regulation, that is, a "negative feedback" between economic development and the environmental changes it causes. In a complex system, feedback mechanisms exist when the state of the system influences its rate of change and then determines a new state. A negative feedback attenuates the changes of a system and thus ensures its stability. For example, when an animal population increases, it will eventually exhaust the food supply in its environment, which increases the mortality rate of that population, and stabilizes it. Similarly, does the worldwide spread of emerging infectious diseases have the potential to cause economic growth to slow down, moderate the rate of evolution of the global economic system, and stabilize it, thus preventing it from getting out of control and having a destructive effect on nature?

The expansionist and dominating tendencies of the human species with regard to natural ecosystems and animal species have favored a growing interaction between humans and microbes, the smallest members of the living world. When these infect people in untouched natural environments, the effect sometimes calms the human colonizing impulse, making microbes the effective "guardians" of nature. For humans, microbes are the most dangerous elements of the entire living world. Indeed, they have a large capacity for biological adaptation, they easily become parasites of living organisms, and are present in every corner of the planet. They are of course much smaller than humans, but there are several billion more of them and they multiply a billion times more rapidly. We live surrounded and constantly covered by these microorganisms, which, luckily for us, are most often not pathogenic. We estimate that in an adult person, 1.5 kilograms (3.3 pounds) of body mass are composed of microbes. Many of them are quite useful to us, too: without them we wouldn't have cheese, wine, bread, or penicillin, nor would we be able to digest food or have healthy skin.

At the origin of civilization, changes in the environment favored the appearance of new diseases, which had a negative impact on societies and which, in some cases, seriously threatened the viability of new modes of subsistence and new forms of social organization. In the Neolithic period around 11,000 years ago, which marked the beginning of the domestication of plants and animals, the change in the food of new populations of farmers, who went from an essentially meat-based diet to a predominately vegetarian diet, led to anemia linked to iron deficiencies, a greater porosity of the bones, and dental problems, such as cavities. A reduction in the size of the jaw linked to chewing more tender food has also been noted by archeologists. Prehistoric humans who lived off hunting and gathering were taller and more robust than their descendants who converted to agriculture. A comparison of various skeletons found in the eastern region of the Mediterranean basin has shown that the size of farmers at the beginning of the Bronze Age was 12 centimeters (4.7 inches) smaller than that of their hunting ancestors of the Paleolithic period. In the West, it has taken the last two centuries of improved diets for humans to reach their previous size (as we saw in the preceding chapter, modern humans in rich countries today suffer from too much meat in their diets, but within a dietary and health context that is quite different from the one experienced by Paleolithic humans).

The small amount of historical data available on life expectancy show that it must have declined from the age of 33 years in the late Paleolithic to around 20 years in Neolithic societies. It then stagnated at around 24 from Roman Egypt (AD 33 to 258) to eighteenth-century France. In 1900, the average life expectancy was still only 26 in regions outside western European countries, countries of European immigration, and Japan. It has only been in the last century that life expectancy has increased rapidly on a global scale. The mortality of children below the age of 15 years was probably higher in Europe in the eighteenth century and the beginning of the nineteenth (one out of two children probably did not reach that birthday) than in prehistoric societies that lived off hunting and gathering. Modern medicine and the great technological revolutions of the last two centuries have reversed this tendency: the increase in life expectancy since the industrial revolution is staggering.

The first farmers quickly understood that to maintain the fertility of the farmland, they needed to spread animal fertilizer and human excrement over the fields. This practice favored the transmission of infectious diseases by allowing the pathogenic agents to multiply in the soil in very close proximity to homes, to percolate in the water, and to contaminate food. The introduction of agriculture in West Africa, 4,000–5,000 years ago, was probably related to the origin of malaria, particularly its most lethal form, *Plasmodium falciparum*. Indeed, the clearing of land and the creation of water sources near a fixed dwelling place created conditions favorable to mosquitoes. To propagate malaria, the same female mosquito must first bite an infected person, then, after a few days while the parasite is developing in the mosquito's digestive tract, bite another person susceptible of being infected. Depending on the species of malaria parasite (*Plasmodium*) and the ambient temperature, from 7 to 35 days are necessary between the ingestion by the female mosquito of infected human blood and the appearance in its salivary glands of new parasites able to be retransmitted to other potential victims. Since a mosquito travels only short distances, it cannot follow a nomadic population, and thus does not contribute to the transmission of the illness among such a population. By contrast, sedentism enables an infected mosquito to retransmit the illness to other people of the same group and thus to amplify the cycle of transmission. It was only after the nomadic way of life was abandoned that malaria became a significant cause of death.

The development of irrigation also created an ideal habitat for mollusks who are vectors of schistosomiasis (or bilharziasis), an intestinal or urinary illness caused by a parasite affecting people who work with bare legs in the water of rice paddies.

The domestication of animals has not only been at the origin of many human diseases (we will look below at zoonoses, or diseases linked to pathogenic agents present in an animal organism and transmissible to humans) but has also favored the circulation of parasites and vectors of diseases between natural environments and the human organism. The presence of a large number of domesticated animals in close proximity to human habitations has facilitated, for example, the transfer of ticks to humans, the transmission of rabies by dogs, and the transmission of toxoplasmosis by cats, cows, and sheep. The concentration of domesticated animals on fixed pastureland has enabled the emergence and the spread to humans of diseases such as tetanus. The animals of a sedentary population graze permanently on pastureland that they themselves have infected with parasites through their excrement, and so the animals are constantly being reinfected, with an increased risk of a transmission of pathogenic agents to humans.

We can see that the domestication of plants and animals, which enabled sedentism and the development of increasingly complex societies, has come at a very high cost to human health. In the long term, however, that strategy has proved to be beneficial to humans. The benefits of agriculture over time in terms of food security, demographic growth, the specialization of work, and the production of an agricultural surplus to meet the needs of an urbanized elite have largely outweighed the costs in terms of health. If our ancestors had not persevered in spite of the heavy burden imposed by new diseases that came out of the Neolithic revolution, we would still be nomadic tribes of primitive hunters with a life expectancy of around 33 years.

INFECTIOUS DISEASES

Even today, we are not free from the threat of microbes. Infectious diseases are at the origin of about one-fourth of the mortality and morbidity of the world's population. They cost the global economy a sum that is measured in billions of dollars each year. These diseases are provoked when there is a notable alteration in the functioning of an organism caused by a microorganism: a virus, a bacterium, a parasite, a protozoan, or a prion. The term *microbe* is a generic term designating bacteria, fungi, protozoa, and viruses, whether they are pathogenic or not. Viruses are biological entities that live in parasites within host cells and are made up of DNA or RNA with a protein coat. Bacteria are living single-cell organisms characterized by an absence of a nucleus (they are thus prokaryote organisms) and of organelles, for example the spirochete responsible for Lyme disease. Parasites are living organisms that feed on, are sheltered by, or reproduce in close connection with another organism, called the host. Protozoa are single-celled living organisms that have a true cellular nucleus (these are eukaryotic organisms), for example the amoeba or *Plasmodium* responsible for malaria. Finally, prions are infectious agents composed of protein in a misfolded form.

In the struggle between humans and pathogenic agents, the latter have a decisive advantage: they benefit from a great capacity to change themselves genetically. Thanks to their numbers, their very rapid rate of reproduction, the frequent modifications in their genetic code, and their ability to adapt to new environmental conditions, they always have a head start in their Darwinian competition with humans. For microbes, a generation is measured in minutes rather than decades. The champion of all categories of genetic plasticity is the virus responsible for AIDS. Its rate of genetic mutation is 10 million times higher than the rate of human DNA mutation. It is therefore very difficult to attack it frontally, because it forms a moving target. An environment characterized by accelerated environmental changes only increases pathogenic agents' advantage over humans: they are in fact much better equipped than the human species to respond to change and to adapt to it very quickly. By modifying ecosystems rapidly and profoundly, humans have thus imposed an additional handicap on themselves in the struggle against microbes.

In nature we live among more than 1,400 known pathogens, which are at the origin of well-identified human diseases. Between 60 and 70% of them are of animal origin. Among emerging diseases of animal origin, 72% are caused by pathogenic agents linked to wild animals, whereas the others are associated with domesticated animals. These zoonoses (from the Greek *zôon*, "animal," and *nosos*, "illness") are transmitted by viruses, bacteria, or parasites present in an animal organism that have adapted to the human organism, thanks to genetic mutations. These microbes or parasites succeed in carrying out a "biological leap" from animal to human, the "bridge" between these organisms of different species being provided by an increased proximity between humans and animals or by a close contact with their excrements and bodily fluids. This is only possible after many fruitless attempts which nonetheless ultimately succeed when there is a great deal of circulation of microbes among animal and human organisms. The more frequent the exchanges of microbes between species, the more opportunities there are to find a genetic mutation that works.

Through their intensive farming of domesticated animals and increased frequenting of wild animal habitats during the last few decades, humans have considerably increased the opportunities for such mutations to occur. In certain cases, a given microbe becomes a specialist of the human organism and can thus be transmitted from human to human. The risk of epidemic is then very high, as in the case of the flu or measles. Fortunately, only 25% of the pathogens of animal origin have acquired that ability to be transmitted from person to person. In other cases, the pathogen remains adapted both to the original animal species and to the human and is not transmitted directly from human to human: it must be constantly reintroduced into a human from animal reservoirs. This is the case, for example, with rabies.

Originally, measles came from cattle; flu from pigs and ducks; whooping cough from pigs and dogs; Lassa fever from rodents; AIDS from chimpanzees; SARS from bats via civets; Ebola fever from bats; rabies from bats, foxes, and dogs. More recently, the H5N1 virus of the bird flu came from ducks and other aquatic and marine birds.

The most deadly viral zoonoses, such as the Ebola virus, do not last long within human populations, unless they are reintroduced frequently from a non-human reservoir. Indeed, between 50 and 90% of people infected by the virus die quickly, inhibiting a large-scale spread of the virus. The strategy of this illness, then, is to be highly contagious from human to human and to be maintained in an animal reservoir, bats, which are little affected by the pathogenic agent. Monitoring animal populations, which can enable the early detection of an emergence of zoonoses, is particularly difficult when the infection of the host animal reservoir is asymptomatic: the animal carries the pathogenic agent in its organism, but shows no symptoms of the illness; thus, it cannot be seen to present a risk of transmission to humans.

Before the very rapid growth of the human population that occurred during the last two centuries, and particularly in the twentieth century, there was a clear segregation between inhabited and cultivated zones, on the one hand, and natural ecosystems, on the other. Today, human activity has colonized all natural environments, which has created a "continuum" between wild animals, domesticated animals, and human populations. Environmental changes thus increase the risk of the emergence of zoonoses.

GIFTS FROM ASIA

Asia, one of the first regions to develop intensive farming and have high population densities, is the likely cradle of many zoonoses coming from domesticated animals, whereas Africa, with its vast forest land with very few humans, has been the source of many zoonoses originating in the wild fauna, including vector-borne diseases (those necessitating the action of a vector—mosquito, tick, fly . . .—to be transmitted: malaria, dengue, sleeping sickness . . .), which we will look at in the following chapter. Between 2000 and 2005, 50 million people were affected by zoonoses, and 78,000 died of them—a figure that still remains modest if it is compared with the tens of millions of deaths every year linked to classic infectious diseases. Some of these zoonoses come from China, Thailand, and Indonesia, where the population is dense and people live in extremely close contact with domesticated animals. During the last 150 years, southern China produced one pandemic of plague, two of flu, and one of SARS.

Several flu epidemics seem to have originated in traditional agricultural environments found predominately in China, which combine the farming of ducks, chickens, pigs, and fish in regions with very dense human populations. According to this hypothesis, a close association between a flu reservoir (a duck) and an environment within which new strains of flu (pig) and an organism likely to be infected (human) intermingle forms an ideal natural laboratory in which to create new epidemics: the birds eat the leftovers from the pigs; the duck excrement fertilizes the ponds where fish are raised; humans eat these three species and take care of them, are in contact with their excrement, and drink water they have contaminated. Different strains of the flu mix together and are combined within the organism of pigs, which is a bit like a casserole in which the flavors of various ingredients are combined to form the final dish. The pig is indeed receptive to the influenza virus of human, avian, and porcine origin. The viruses circulate almost freely among these various species through feces-mouth contact, until they form a highly contagious strain of flu, which then colonizes the world in a few months via the international transportation system. In the case of the bird flu, chicken constitutes an intermediary stage in the transmission of the virus to humans. In Asia, the highest number of cases of chickens infected by the H5N1 virus of bird flu is found in regions with rice paddies where flocks of domesticated ducks are taken to feed after the harvest.

On the Asian continent, animal production for human consumption has increased eightfold in less than thirty years. This farming is concentrated on large industrial farms outside the cities, which creates ideal conditions for the emergence and the propagation of pathogenic strains. Indeed, in those farms, the domesticated animals have an immune system that is greatly weakened by their cramped living conditions. The high densities of human populations near these farms, with poor neighborhoods and dubious hygiene conditions, favor the transmission of infectious agents to humans through the liquid effluvia from the farms, which get mixed with drinking water. The worldwide exportation of animal products, from Thailand above all, enables the spread of pathogenic agents through transnational trade routes. East Asia is the great petri dish of the planet; it produces then disseminates new microbes throughout the world almost annually.

VULNERABILITY

The health of human populations depends largely on natural and socioeconomic environments. Around forty years ago, the industrial production of antibiotics, vaccines, and insecticides, along with public health policies, gave the illusion that humans had conquered infectious diseases. In 1969, William H. Stewart, then Surgeon General of the United States, is alleged to have declared that the time had come to "close the book on infectious diseases."[3] A few decades ago, few scientists had been aware that at the same time that "classic" diseases were retreating, at least in industrialized countries, new diseases were appearing at a steady rate. The world of microbes is far from stable. It has become

3. Although this remark has often been cited in the literature, the source of the statement could never be located, and whether Dr. Stewart actually made such a comment could never be confirmed.

clear today that environmental changes and globalization have created multiple opportunities for the appearance of new virulent pathogenic agents. These are all the more dangerous since the grip of emerging diseases has become global, and humans have not developed immunity in the face of microbes that were previously unknown.

Diseases do not strike societies randomly any more than they are divine punishment: they emerge from relationships that human beings create among themselves, with the animal world, and with the natural and urban environments. Pathogenic agents take advantage of the slightest weakness to infiltrate human organisms in places where poverty, a lack of hygiene, a change in the environment, and a concentration of animals in unusual conditions or locations have created opportunities for infection. The emergence of new infectious diseases is linked to several forms of environmental change: changes in the physical environment, connected both to the climate and to the vegetation, and to land use; urbanization, which concentrates a critical mass of people susceptible to infection; and intensive farming, which increases close contact between humans and domesticated animals. A fourth factor, the increasing mobility of human and animal populations, favors the global dissemination of these diseases once they have appeared.

The poorest societies are also the most vulnerable in the face of emerging diseases. Infectious diseases are the cause of 1 to 2% of deaths in the richest countries, and of 50% of deaths in the poorest countries. Thus, for example, once its phase of worldwide diffusion passed, AIDS today affects mainly sub-Saharan Africa, where the absence of food safety, bad hygiene conditions, a low level of education among the population, and ineffective health services facilitate the appearance of new diseases. Poverty and a poor education system are the main obstacles to an improvement in health.

One of the principal effects of environmental change is the increase in disparities between rich and poor countries, notably in matters concerning health. There is a large gap on a global scale between vulnerability in the face of new diseases and financial means devoted to public health. Most of the emerging diseases originate in developing countries, notably in South Asia and Southeast Asia, whereas most of the world resources for public health, the surveillance of emerging diseases, and biomedical research on new vaccines and medication are concentrated in the rich countries. Three figures concerning Africa illustrate this gap: that continent contains 10% of the world's population, 24% of the world burden of diseases, and only 3% of the medical personnel on the planet. Recently, Bill Gates noted that we devote more money to research for a treatment against baldness than we do for the eradication of malaria in the world.[4] To decrease human vulnerability in the face of emerging diseases requires above all a reduction in poverty, but also an increase in the ability to respond rapidly to new epidemics in regions where they are most likely to arise. For this we must anticipate the threats to come, through a better understanding of factors that cause their emergence. In an editorial in the journal *Science* published in 2006, a group of scientists identified eight categories of infectious diseases that are likely to appear in the near future:

1. New emerging diseases such as SARS, or novel variants of existing diseases such as those of the H5N1 virus of bird flu;
2. Infections that have developed a resistance to treatment, such as bacterial infections resistant to antibiotics (the very dangerous methicillin-resistant staphylococcus aureus, or MRSA);
3. Zoonoses such as Lyme disease, anthrax, or infections linked to food (*Escherichia coli* O157:H7, salmonella);
4. The three big tropical diseases that chiefly affect the developing countries and that pursue their slow but deadly progress: AIDS, tuberculosis, and malaria;
5. Plant diseases affecting crops and thus agricultural production and food security, especially in poor tropical regions;
6. Acute respiratory, viral, or bacterial infections, which threaten to become more serious through climate change and urban pollution;
7. Sexually transmitted diseases;
8. Animal diseases such as foot-and-mouth disease or mad cow disease, whose main economic impact is to create commercial barriers to the importing of animal products.

Most of these health risks are amplified by environmental changes in different forms: modification of animal habitats that are the source, the hosts, or the vectors of pathogenic agents; changes in farming practices; urbanization; and climate change. The nineteenth and twentieth centuries experienced extraordinary progress in the realm of medicine, from discoveries of vaccinations and antibiotics to a scientific approach to public health and hygiene. We must recognize, at the beginning of the twenty-first century, that economic development comes

---

4. In a speech to University of Chicago students, February 2008; see also the 2009 Technology Entertainment Design (TED) conference, online at http://www.ted.com/talks/lang/eng/bill_gates_unplugged.html.

at a cost to the environment and to our health; that knowledge must become an important factor in both political and individual decisions. Whereas humanity is endowed with a remarkably efficient arsenal to fight emerging and reemerging diseases, the current mode of development and its impacts on the natural environment have also increased the frequency with which such diseases now appear.

THE THREAT OF ARTHROPODS

If the organisms most dangerous to humans are not large carnivores—lions, tigers, and wolves—but pathogenic microorganisms, the danger those microorganisms represent becomes even greater when they ally themselves with another branch of the living world—arthropods (for example insects, myriapods, arachnids, crustaceans). This strategic alliance renders vector-borne diseases particularly threatening to humans. The risk of their transmission is also closely associated with environmental conditions. Several of the diseases transmitted by mosquitoes (such as dengue, West Nile virus, Chikungunya, Rift Valley Fever) and by ticks (such as Lyme disease, tick-borne encephalitis, Crimean-Congo hemorrhagic fever, anaplasmosis) are spreading rapidly in many regions of the world.

In this chapter, I will discuss some theories that explain this expansion. Is it due to climate change, a view that seems to have now become commonplace? Or is the upsurge the result of a transformation of natural ecosystems through human activity: deforestation, agricultural expansion, and the fragmentation of landscapes? Or should we blame socioeconomic changes and a modification of human behavior? Or perhaps these epidemics are following the natural course of biological evolution? As we will see, the responses involve a combination of these various factors. Unfortunately, it's not possible to take any simplistic shortcuts to explain the emergence of these vector-borne diseases. (The reader who wishes to avoid the slightly more technical analysis in this chapter may pass directly to its conclusion and continue reading without losing the thread of the book.)

Arthropods are invertebrates that represent 80% of known species in the animal kingdom, with more than 1.5 million different species in their phylum. Several parasitic arthropods, stingers or suckers, are vectors of diseases called vector-borne diseases: the infectious agent is transmitted from one infected individual to another through an invertebrate host called a vector, such as certain mosquitoes or flies, hematophagus acarids (those who feed on blood) such as ticks, and ectoparasitic insects (external parasites) such as fleas. The pathogenic agents have "signed" an associative contract with the arthropods that is often mutually beneficial: the arthropods shelter the agents in their organism, transport them, and introduce them into the body of their victim, in exchange for which some of those microorganisms provide services to their hosts, including producing amino acids or vitamins, a vital functional role; providing a metabolic complement in situations of environmental stress; and changing the makeup of the host, giving it a certain immunity. Neither party wants the pathogenic agent to decrease the chances of survival of the arthropod who is housing it: for the arthropod, the reason is obvious; as for the microbe, it would lose its chances to multiply.

For human vector-borne diseases, four agents are involved in the cycle:

1. The infectious agent: virus, protozoa, bacterium . . . ;
2. The invertebrate vector: fly, mosquito, tick . . . ;
3. The intermediate host: wild or domesticated animals, which are a reservoir for the infectious agent;
4. The human host, who falls ill once infected.

The infectious agents circulate essentially among their animal hosts and their vectors; but occasionally they escape toward humans and infect them. Whereas infectious diseases are responsible for 15 million deaths every year, or 26% of the total number of deaths, the subcategory of vector-borne diseases causes 1.43 million deaths each year. Those vector-borne diseases transmitted by mosquitoes, mainly malaria, are responsible for most of that gloomy statistic, with 1.3 million deaths per year. The morbidity that is associated with them (that is, the number of people afflicted in a population) is also very great.

These diseases are particularly sensitive to environmental changes, because the conditions of the environment influence the life cycle not only and especially of the vector but also of the intermediary hosts. The arthropod vectors do not have highly developed physiological mechanisms to regulate their temperature and their body humidity. The rate of their reproduction, their mortality rate, the frequency with which they feed on the blood of mammals, and their ability to transmit a pathogenic agent thus depend strongly on the ambient temperature and its seasonal variations, the humidity at ground level, and the presence of habitats enabling reproduction and the survival of each vector in the vegetation, the soil, and the topography. Furthermore, the circulation of the infection often depends on the presence of an intermediary host, that is, those animals, mammals or birds, which are reservoirs of the infectious agent and thus enable the cycle of transmission to be constantly renewed. These animals also have very specific needs in terms of the habitat they require to reproduce and thrive. A change in the natural environment can increase or decrease the population density of these intermediary hosts, which has a determining effect on the risk of disease. Everywhere in the world, humans domesticate nature, alter the landscape, transform natural environments into agricultural zones, and modify land to derive the greatest production from it with the help of irrigation; the massive application of fertilizer, herbicides, and pesticides on the fields; and control over forests by selecting species, draining swamps, and so on. These transformations change the habitat of vectors and animal hosts, and thus the entire epidemiological environment.

Close to 3,500 species of mosquitoes have been identified, only a small fraction of which are vectors capable of transmitting disease. Each species has its specific habitat preferences for its reproduction and survival in the larval and adult stages; they all have their own ecological niche. The development of a mosquito's eggs requires the presence of stagnant, warm, and often shallow water. The presence of irrigated farmland, hydroelectric dams, pools for fish farming, ditches formed by passing vehicles, or simply receptacles that contain water increase the capacity of an environment to maintain a large population of mosquitoes. Among the 60 or so species that transmit malaria, some live in forests. This is the case, for example, of *Anopheles dirus*, a vector of malaria in Southeast Asia. Deforestation will thus push the disease back. However, other species that are also vectors of malaria prefer the edges of forests. This is the case of *Anopheles darlingi*, the main vector of malaria in the Amazon basin. In that case, deforestation accompanied by a fragmentation of the forest canopy which then forms small forest patches, or the construction of a road through a forest, increases the total length of the forest edge. The increase in the number of habitats favorable to this mosquito enables an increase in its population, which further increases the risk of transmission of malaria among the farmers who live on the edge of forests. And other species, such as the African *Anopheles gambiae*, develop better where the forest has disappeared and where it is hot and sunny.

Deforestation will thus be associated with new epidemics of malaria. To complicate things further, mosquitoes adapt genetically within a few generations to new environmental conditions, including the use of new insecticides. Their preferences thus evolve along with environmental changes. As soon as a zone has been abandoned by a species of mosquito following an ecological modification, another species immediately comes to colonize it.

In natural conditions, yellow fever (also known by the nice name of *vomito negro*, which requires no translation) is transmitted by *Aedes africanus* mosquitoes to many species of monkeys who live on the canopy of tropical forests in South America and Africa. In Africa, where the disease originates, these monkeys are almost all infected, but they don't suffer from the disease. By contrast, several species of monkeys in the New World are very sensitive to infection and have a high mortality rate from it. Since deforestation has caused tropical forests to recede, monkeys and mosquitoes have adapted to a new way of life, on ground level, where they have found food sources linked to the presence of humans. The *Aedes simpsoni* mosquito, which feeds from monkeys and humans and has thus enabled the virus to circulate among these two species, is at the origin of the human form of yellow fever. Today, other species of mosquitoes also contribute to the transmission of the disease. Wherever wild fauna is diminishing due to hunting and the destruction of natural habitats, mosquitoes must either retreat, or adapt by taking a greater proportion of their meals from human beings. This last strategy favors epidemics and the appearance of new vector-borne zoonoses.

Another mosquito of the *Aedes* genus, *Aedes aegypti*, transmits dengue, a deadly disease spread widely in the tropics. It reproduces in man-made containers that hold water—a dog's water bowl, a barrel that collects rainwater, vases in cemeteries, or blocked gutters—and in wells. Urbanization increases the numbers of these containers and thus the mosquito population. Dengue has therefore become an urban and periurban disease, which explains why there are so many victims in regions where cities are growing rapidly. This phenomenon reflects the very rapid adaptation of that mosquito, which originates in the forests of Africa, to new urban environmental conditions—a beneficial strategy from the mosquito's point of view, in a world where more than half the global population now lives in cities. *Aedes aegypti* has by chance acquired suitable biological attributes to multiply in cities: it has benefited either from good adaptive flexibility, or from preexisting skills that have enabled it to adapt easily to an urban way of life and reproduction. This mosquito, also able to transmit yellow fever, is thus responsible for urban epidemics of that disease. Because of *Aedes aegypti*, the yellow fever virus has seen the population of its potential human hosts increase considerably. The *Culex* mosquito, one of the vectors of the West Nile virus, reproduces in underground storm drainage systems found in all large cities.

Socioeconomic conditions and human behaviors also play an essential role in the threat of vector-borne diseases, as is proved by a study on dengue carried out at the border between Mexico and the United States. Two communities, on either side of the border, were followed for a season. The climate conditions were the same, but the *Aedes aegypti* vector was more abundant on the American side of the border. However, on the Mexican side, a clearly higher number of people had been infected. This difference was explained by the higher number of air-conditioning systems in American houses. The lower interior temperature was the best protection against mosquito bites.

MALARIA AND CLIMATE CHANGE

Has climate change already—or will it soon—increase the risk of infectious diseases, notably those transmitted by arthropod vectors? This question is the subject of impassioned debate among scientists, even if everyone agrees that climate change has an undeniable influence on vector-borne diseases. Whereas some theories tend to simplify the issue, others reveal such complexity that any prediction becomes impossible.

Let's take the example of malaria, which is transmitted by several species of mosquitoes of the *Anopheles* genus. Climate change threatens to cause this mosquito to appear in regions that at present are a bit too cold for it to be in its comfort zone. By contrast, it might disappear from regions whose climate will become too hot and too dry. A dry climate, in particular, greatly limits a mosquito's ability to survive, because small, damp habitats in which it reproduces become rarer. Where mosquitoes are already present, a rise in temperature—but up to a certain threshold—increases the frequency with which female mosquitoes feed on the blood of the host, human or not. A warming of the climate also accelerates the biological development cycle of the parasite in the digestive system and the salivary glands of the mosquito: it could go from 20 to 10 days. These two effects thus increase the risk of malaria. But warming also diminishes the mosquito's life expectancy, which will go from around 20 days to 10. For a malarial transmission risk to exist in a region it is necessary, among other things, for the climate conditions to be such that the biological development of the parasite in the mosquito's organism be shorter than the life expectancy of the latter. Furthermore, in regions where exposed people already receive many bites from infected mosquitoes (up to 300 per year), the incidence of malaria will not increase with global warming: once the infection is transmitted, additional bites won't change anything.

As we have seen, the transmission of malaria can occur only under very particular climate conditions, and the transmission cycle is very fragile. For the malarial parasite *Plasmodium falciparum*, which is the most common and the most deadly, the lower temperature limit for transmission oscillates between 16° and 19°C (60.8° and 66.2°F) and the upper limit between 33° and 39°C (91.4° and 102.2°F), whereas *Plasmodium vivax* has a somewhat lower temperature limit, 14.5°–15°C (58.1°–59°F). A slight change in the climate can cause a geographic region to escape that climate envelope, or on the contrary cause it to enter it. Some species of mosquitoes are, however, very resistant to extreme temperatures. *Anopheles arabiensis*, a vector of malaria in Africa, survives in the thatch roofs of houses when the outside temperature climbs to 55°C (131°F). In the past, in Lapland in northern Scandinavia, mosquitoes spent the glacial winter sheltered in stables or houses, sometimes biting hosts, to whom they transmitted malaria.

The spatial distribution of the disease can shift without there necessarily being an increase in the total number of victims on a global scale. In fact, there could be as many regions from which malaria will disappear as regions where it will appear. It is nonetheless probable that the former will be less populated than the latter, which will then lead to an increase in the total number of cases of malaria in the world. Indeed, in Africa, semiarid regions, which could become too hot and too dry for the vectors of malaria (Sahel, for example), have lower population densities than the higher zones (the high plateaus in Ethiopia, for example), which global warming could make more favorable to mosquitoes. Throughout history, populations have occupied these plateau regions precisely because the risk of the disease was lower. In places where malaria will appear, populations will not yet have developed immunity, and the risk of an epidemic will be higher. Furthermore, an aridification of the climate might lead to an expansion of irrigated farmland, which will create habitats favorable to mosquitoes throughout the year. Still, all of this remains quite hypothetical, and it is difficult to take into account all the factors that come into play. For example, depending on whether the maximum or minimum daily temperature will increase more or less rapidly in a place, the impacts on the risk of transmission of vector-borne diseases will be different.

Even if the geographic range should expand for mosquitoes who are vectors of malaria, that would not necessarily mean that malaria would have more victims: an effective system of public health and preventive measures could prevent the parasite responsible for the disease, and for which the mosquito is only a means of transport, from propagating within a population. A good early diagnosis of the infection, access to appropriate medicine, and a systematic use of protective means against mosquito bites can prevent the spread of the disease. At the end of the nineteenth century in northern Europe, malaria was rampant in the summer, even in the south of some Scandinavian countries, and including Great Britain and especially the Netherlands. It was only in 1975 that the last pocket of malaria endemic to Europe (in the Macedonia region of Greece) was eliminated. The eradication of the disease was linked not to a cooling of the climate—on the contrary, it was getting warmer at the time—but to three other factors: the drainage and cultivation of swampy lands, and thus the elimination of habitats where mosquitoes reproduced; an increase in the population of livestock, notably in stables close to homes, which attracted mosquitoes to that alternative food source and thus decreased the number of bites inflicted on humans (let's note the positive role played here by a close contact with animals); and the overall improvement in living conditions, hygiene, and health policies within the human populations.

Once the role of mosquitoes in the transmission of malaria was discovered, the application of an insecticide like DDT contributed to the control of the disease where it still existed, at least until mosquitoes developed a resistance to DDT through genetic mutation. This resistance has contributed to the reemergence of malaria beginning in the 1970s, and which continues today.

In southern regions of the United States and Europe, in the Camargue, for example, one finds mosquitoes that are competent vectors of malaria. From time to time, people who become infected during a trip to a tropical region, and who thus carry the malaria parasite in their blood, reside in regions where climate conditions are favorable to transmission. And yet, cases of a local transmission of malaria are very rare, even though they do exist. It is quite simply because, in these countries when new diseases emerge, the public health system enables early detection, rapid intervention, and prevention.

Malaria is above all a disease of poverty. Its resurgence in Africa is explained less by climate change than by the growth of the population around swampy zones where mosquitoes reproduce; the expansion of irrigated rice-growing; deforestation; the migration of unimmunized people to regions at risk; an increase in the population of refugees who, fleeing conflicts, are regrouped in unhealthy zones and live in frightening hygiene conditions; the disorganization of prevention systems for lack of adequate financing for mosquito eradication operations; and the appearance of parasites that have become resistant to successive medication against malaria, as well as mosquitoes that have become resistant to insecticides.

Regarding diseases transmitted by ticks, what is responsible for their emergence: climate change, the transformation of ecosystems by humans, socioeconomic changes, or biological evolution? The story becomes even more complicated here than for the diseases transmitted by mosquitoes. Act like a tick on the skin of its victim: stick with us for a moment.

In Europe, the diseases that ticks communicate to us have been increasing rapidly during the last two decades, in particular Lyme disease (*Lyme borreliosis*) and tick-borne encephalitis. The first of these diseases was identified in 1976 on the East Coast of the United States, in the village of Lyme, Connecticut, from which it derived its name. In reality, it may have already been present in a diffuse manner in North America before European colonization. Furthermore, its symptoms had been described in 1910 in Europe, and in 1950 they were associated with the bite of the tick *Ixodes ricinus*. Lyme disease is observed today in most temperate regions of the world, with close to 100,000 new cases identified each year. Given that many cases are not diagnosed, the true incidence could be three times higher. This disease is concentrated in high-risk, hyperendemic (that is, exhibiting a high and continued incidence) regions. In these "hot spots," the disease can affect one person out of 1,000 each year or, in extreme cases such as the Lyme region, one person out of 100.

Tick-borne encephalitis is also transmitted by the *Ixodes ricinus* tick, but with a less effective transmission cycle. This disease is fatal in 1% of cases. There were an increasing number of cases in Eastern Europe just after the end of Communist rule: the number of cases increased from twofold to thirtyfold between 1992 and 1993, depending on the country; the Baltic states, Poland, and Slovakia experienced the most rapid growth. Tick-borne encephalitis has emerged slowly in western Europe since 1998 from Sweden to Switzerland, including France and Germany, with a sharp increase in cases in 2006.

Only the interaction of several ecological, climatic, and socioeconomic factors can explain the emergence of Lyme disease and tick-borne encephalitis. The life cycle of ticks goes through three stages: larva, nymph, and adult. At each of these stages, which can spread over a season, a single meal of blood is taken from a host, either animal or human. The larva feed essentially on small animals that move around on the ground. Only the adults, and to a lesser degree the nymphs, can climb high enough onto vegetation to jump onto a deer or a human. Ticks become infected only when feeding. In the case of tick-borne encephalitis, the infectious agent does not survive long in the organism of the host animal. For the transmission cycle to be effective, the larvae and nymphs must have a synchronized activity: they must look for a meal at the same time and feed together on the same rodent. (In the case of Lyme disease, where the infectious agent survives a bit longer in the host's organism, there can be a slight delay.) The larvae then catch the infectious agent via the blood of the mammal and can retransmit it to humans during their next meal, once they have become nymphs or adults. Most people are infected by nymphs, because they are small and difficult to detect. This mode of transmission of the infection from an infected tick to several larvae feeding on the same host enables an amplification of the microbe's presence in the ecosystem. Despite the inevitable mortality of a high proportion of ticks, this enables maintenance of the circulation of the infectious agent within the population.

The biological activity of nymphs begins at a lower spring temperature than that of larvae: nymphs begin to be active at the end of winter, as soon as the temperature rises above around 7°C (44.6°F), and the larvae as soon as it goes above 10°C (50°F). If spring comes on quickly, with a sharp rise in temperature, the probability that the nymphs and larvae become fixed on the same host at the same time is higher. If at least one of the nymphs is already infected, there is then a greater chance of the infection's presence in the tick population: a large number of larvae will then be carriers the next season, once they have become nymphs. The risk of infection is thus closely linked in part to the climate, but more to the speed of transition from winter to spring than to a simple increase in average annual temperature. Thus, to directly link global warming to the appearance of the disease is too simplistic, especially since other biological factors render the situation even more complicated.

When there is an increase in the animal population on which ticks feed, the population of ticks also increases the following year, because the number of female ticks that produce eggs increases. All the animal hosts are not, however, able to retransmit the infection from one tick to another. For Lyme disease, rabbits, foxes, badgers, weasels, ermines, skunks, and lizards do not reinfect the ticks that feed on their blood, even when they are themselves infected. By contrast, field mice, voles, and shrews are very effective in this retransmission of the pathogenic agent. The greater the diversity of animal species, the more the infected bites of ticks threaten to be diluted in ineffective species: a higher proportion of bites from infected ticks is then lost in a biological "dead-end," since the pathogenic agent is not put back in circulation with subsequent bites. It is a bit as if, during a soccer or basketball match, half the players made the ball disappear each time they received a pass. We can be sure that at the end of the match there would be fewer points scored than if all the players moved the ball around.

Some ecosystems are greatly impoverished in terms of biodiversity. A transformation of the landscape by humans and a fragmentation of natural habitats cause the disappearance of the most demanding species in ecological terms, while very adaptable and generalized species, such as small rodents, will hardly be affected. Their population will even increase, insofar as their predators and the species with which they compete for food will have disappeared. Some of those species are precisely those that are the most effective for the transmission of the pathogenic agent to ticks. A concentration of most of the tick bites onto effective rodents thus increases the infection by retransmitting it to all other ticks who feed on their blood. Animal biodiversity thus renders a precious service to us by diminishing the proportion of infected ticks in our environment. This mechanism of dilution associated with increased biodiversity in the community of hosts was proposed in 2009 by Richard S. Ostfeld, an American biologist.

In reality, the ecology of diseases transmitted by ticks proves to be even more complicated than this mechanism of dilution. Indeed, the animal species not effective in transmitting the infection can nevertheless contribute to an increase in that infection. This is above all the case with deer and other members of the deer family, whose population has sharply increased in Europe and the United States during the twentieth century following the quasi-eradication of their natural predators. They are the favored source of food for adult ticks, because their great mobility in forests frequently enables ticks perched on ferns or tall grass to latch onto their bodies as they go by. A deer can carry and feed a large population of ticks. Adult female ticks to which an easy source of food is thus offered produce a large number of eggs, which enables ticks to proliferate the following year. In subsequent seasons, a certain proportion of new larvae and nymphs will end up on infected rodents and will infect other larvae. In these cases, a species, the deer, which is supposed to dilute the infection, in fact contributes to its dissemination by enabling an increase in the population of ticks that circulate in the ecosystem. This phenomenon would explain, for example, the recent increase in cases of tick-borne encephalitis in certain regions of the Italian Alps where the deer population has increased more than tenfold since the 1950s.

There is another element of complexity: a sharp decline in the fauna, and thus of potential hosts for ticks, will increase the number of ticks attached to vegetation awaiting a host on which they can have a meal, which will increase the risk that a human being will be bitten. By contrast, that risk will be relatively small if the animal hosts are very abundant.

For there to be a transmission of the infection to a human, it is necessary for the infected ticks looking for a meal to encounter an unvaccinated human (there is a vaccine against tick-borne encephalitis, but not against Lyme disease). The incidence of these diseases increases then with the human population that frequents forests, notably for recreational purposes: a Sunday stroll, a bike ride, jogging, camping, gardening near a forest, and so on. In some countries, Lyme disease is very prevalent in wealthy suburban zones where isolated houses with large yards border forests and whose inhabitants enjoy many leisure activities in the forest.

The painstaking work of Sarah E. Randolph, a biologist at Oxford University, has enabled us to understand the rapid increase in the number of cases of tick-borne encephalitis in Eastern Europe in the 1990s and in central Europe in 2006. Following the fall of communism in Eastern Europe, many people with low-paying jobs, who had lost their jobs in collective farms or public enterprises, were infected by tick-borne encephalitis by frequenting the forests in which they gathered what they could to survive: mushrooms, wild berries, wood for heating, and so on. In the Baltic states, for example, close to half the population was still unemployed in 1999. Picking edibles from the forests had become an important source of food for these poor social classes. In addition, as the agricultural land of former collective farms had been abandoned, it was invaded by shrubs and tall weeds that formed a habitat favorable to rodents. A decrease in the use of pesticides also encouraged the growth of the rodent population. This explains in part the explosion of cases of tick-borne encephalitis beginning in 1993 in former countries of the Soviet Bloc.

In Denmark, the expansion of Lyme disease was correlated with an increase in the densities of deer during the last decades.

In Germany, Switzerland, the Czech Republic, and Slovenia, the number of cases of tick-borne encephalitis increased sharply in 2006, whereas the following year, the incidence of this disease returned to normal levels. The following hypothesis was advanced by many experts: the winter of 2006 was very cold, and the weather during the following summer and autumn was exceptionally hot and dry; those very mild temperatures would have had an effect on the tick population, their rate of activity, and their spread to higher altitudes and latitudes. But the systematic tick-trapping campaigns of biologists in several countries, in 2006 and 2007, led to the conclusion that in reality those exceptional climate conditions did not have a significant influence on the abundance of ticks. Sarah Randolph and her colleagues have shown that the large number of human cases of infection was indeed linked more to an increase in human activities in forests, which the good weather had encouraged: many people took advantage of this long sunny and dry period to enjoy activities in a natural setting. Furthermore, the climate conditions were ideal for the growth of mushrooms and wild fruit. And in the Czech Republic, Poland, and the Baltic states, there is a long tradition of picking mushrooms in the woods, which are moreover exported in large quantities to western Europe.

We must thus beware of simple correlations between the emergence of new diseases and climate change, as they hide complex ecological and land-use mechanisms. It is probable, however, that in the future, in response to even greater climate change, the geographic range of ticks will be extended. This has been observed in the Czech Republic along an altitudinal gradient in a mountain region: at the beginning of the 2000s, at the same time that the temperatures were increasing, ticks became more abundant at a higher altitude than what had been observed at the end of the 1950s. In Sweden, the distribution of ticks expanded to the north between the beginning of the 1980s and the mid-1990s, following a decrease in the number of very cold days in the winter. In Germany, it has been shown that adult ticks and nymphs remained active longer during mild winters. Nevertheless, the incidence of tick-borne diseases will always depend on three factors: an abundance of ticks, the prevalence of the infection in the tick population, and the rate of contact between ticks and humans susceptible to infection. As we have seen, climate is just one of the many variables controlling these factors.

THE ASIAN TIGER IN EUROPE

The occurrence of 200 human cases of Chikungunya, another disease transmitted by mosquitoes, in Europe during the summer of 2007 was attributed somewhat hastily to global warming by the World Health Organization. This first epidemic of a tropical disease in Europe originally appeared in the village of Castiglione in the Ravenna Province of Italy, and then in neighboring villages. It began in 2004 in Kenya and had then stricken 40% of the population of Reunion Island in 2005 and 2006 before moving to India, where it infected close to 1.5 million people. At the origin of this epidemic was a small genetic mutation of the virus, which facilitated its reproduction in the intestines of mosquitoes. Furthermore, the virus adapted to the mosquito of Asian origin, *Aedes albopictus*, a very effective vector also known by the name of Asian Tiger mosquito because of the black and yellow stripes on its body. Before, only *Aedes aegypti* (the vector of dengue and yellow fever discussed above) was a vector of Chikungunya.

Aedes albopictus had already begun to colonize several European countries and the United States twenty years earlier, via the international trade of used tires. We find it today in Mediterranean countries from Greece to Spain and farther north in Europe, as far as Holland, where it survives in greenhouses containing houseplants (lucky bamboo) that come from Asia.

During the summer of 2007, the Chikungunya virus was imported into Italy in the blood of an infected person who had returned from a trip to India. By chance, this person was bitten by a mosquito from the competent species that, a few days later, bit someone else, thereby setting off the cycle of transmission. Most of the victims lived in the lower part of the Italian village of Castiglione, along a river where mosquitoes reproduced. Surprisingly, very old people were spared: since they rarely went outside, they were not exposed much to mosquito bites. The true guilty party of this epidemic was thus not climate change but rather globalization, which first brought larvae of *Aedes albopictus* to Italy in used tires and then, twenty years later, a traveler who the day before had been in India, where he contracted the infection. The work of genetic mutations within the virus and bites on the wrong person at the wrong time did the rest.

SOMETIMES IT REALLY IS THE CLIMATE

Some infectious diseases seem to be linked more directly to the climate, among other factors. This is true of bubonic plague, which each year affects between 1,000 and 3,000 people throughout the world. Each of the three great pandemics of the plague that occurred in history—they will be discussed in the next chapter—were probably influenced by a climate event. Bubonic plague is caused by the *Yersinia pestis* bacterium, transmitted to humans through fleas that feed on and are transported by rodents. Climate conditions influence the three actors: the bacterium, the flea, and the rodents. Several hypotheses have been proposed, and although they have received an empirical confirmation in some regions of the world, they must still be validated. Above a temperature of 27°–28°C (80.6°–82.4°F), the bacterium is no longer retained and transmitted by fleas, which halts plague epidemics. The rate of survival of fleas also decreases with very high temperatures, and that of larvae in very humid air. The ideal climate envelope for the transmission of this disease by fleas thus corresponds to a temperature between 24°C (75.2°F) and 27°C (80.6°F).

Researchers have shown that in central Asia, a greater amount of *Yersinia pestis* bacteria was found in rodents during years when the spring was hot and the summer humid. Fleas become active in the spring, once the temperature goes above 10°C (50°F). A warm spring, without night frost, enables them to reproduce early in the season and to continue to multiply during the summer if there is abundant rain. A large population of fleas increases the transmission of the infection among rodents. A temperature increase of only 1°C (33.8°F) increases by half the bacterium's prevalence among its animal hosts' population.

Finally, the rodent population also influences the risk of transmission. An increase in rain increases the amount of food available to rodents and considerably accelerates their rate of reproduction. Once the density of rodents goes beyond a certain threshold, the transmission of the infection among them becomes much more probable. During the plague pandemics in the Middle Ages, the climate conditions favorable to bubonic plague were common in central Asia, from which the pandemics seem to have originated. Such conditions could become more frequent with climate change. Quite fortunately, the populations of black rats (Rattus rattus), which were abundant in Eurasia in the Middle Ages, have been in part replaced today by the brown or Norway rat (Rattus norvegicus), whose ability to transmit the plague might be weaker.

In another part of the world, the United States, a dozen human cases of bubonic plague are identified each year. But since the 1960s these cases have shifted toward the north of the country, thus accompanying an increase in mean annual temperatures.

At present, a single case of an expanded geographic range of a vector-borne disease has been attributed mainly to climate change (although that remains a hypothesis, questioned by some). It concerns an animal and not a human disease—catarrhal fever, or bluetongue disease, which affects ruminants and above all certain species of sheep. During the summers of 2006 and 2007, an epidemic with major economic impacts spread by surprise to northern Europe from a source in Maastricht. The bluetongue virus is transmitted by certain species of culicoides, biting midges present in tropical and subtropical regions that move over long distances, carried by the wind. Bluetongue disease became established in Mediterranean Europe only in the autumn of 1998. Before that it had appeared only sporadically and briefly, due to favorable winds or the trade of livestock coming from Africa. The simultaneous incursion of several different strains of the disease in southern Europe beginning in 1998 and farther north in 2006 can be explained in part by climate conditions, particularly a significant increase in nocturnal and winter temperatures, with cross-border movements of animals also playing a role. This enabled both a northerly movement of the primary vector of bluetongue disease, Culicoides imicola; the survival of that midge and of the virus during winters that have become milder; and, even more disturbing, the acquired ability to transmit the disease by northern European species of midges (Culicoides obsoletus and Culicoides pulicaris). The traditional vector of the disease thus succeeded in "passing the baton" to other local species, and the progression and long-term establishment of the disease in northern Europe were facilitated. Since its arrival in Benelux and in Germany in 2006, it reached Great Britain and the Czech Republic in 2007 and southern Sweden in the autumn of 2008. Rarely fatal, bluetongue disease nevertheless leads to a high rate of economic losses due to a developmental delay and low productivity in the affected sheep.

Another serious disease, cutaneous leishmaniasis, is transmitted to dogs and to humans by several species of sand flies, of the subfamily Phlebotominae, that are found notably in the Mediterranean basin. It is also present in Asia and East Africa, and affects 12 million people worldwide, with 1.5 million to 2 million new cases each year. Climate change could enable the sand flies to extend their range of influence to northern Europe, but this remains hypothetical, especially since it is unlikely that all the species that are vectors of leishmaniasis will respond to it in the same way. All the same, cases of cutaneous leishmaniasis have been found north of the normal geographic range of the disease. The vector also seems to be evolving toward southeastern Europe. It has been shown, however, that this distribution is also controlled in part by topographical features and by the type of soil. Feral dogs that visit untouched zones and export the parasite into endemic regions also contribute to the geographic expansion of the disease. In a region in Brazil, the number of cases increased following a drought. According to one hypothesis, this increase would be connected less to a change in the distribution of the vectors owing to a decrease in rain than to a modification of human behavior. Indeed, the lack of water in a rural environment forced many people to migrate to the cities and to regroup around water sources, humid places that make up the preferred habitat of sand flies. The risk of transmission of the disease thus increased around the water sources in this region of Brazil.

DISTURBING A COMPLEX SYSTEM

It is clearly not possible to escape complexity when we attempt to understand an increase in the risk of transmission of vector-borne diseases. The causes of their emergence or of a geographic expansion are always multifactorial and so must be sought in the complex interactions between pathogenic agents; their vectors; the animal hosts, whether competent or not; the natural habitats of those vectors and hosts; the climate; and humans. According to the metaphor of an American doctor in the 1930s, Lewis W. Hackett, as in a chess game, there are only six types of pieces, but an infinite variety of situations. These interactions are influenced by changes to the ecosystems, a modification and fragmentation of habitats, and the way the land is used by humans. Small changes in human habits and behavior can have unpredictable effects on the risks of exposure to infections. Climate change plays an important role insofar as it influences those human interactions and behaviors. The complexity of the transmission cycle of these diseases can amplify the impact of a disturbance and render any prediction difficult. Any modification in the equilibrium of an epidemiological system based on the interactions among multiple actors can lead to surprises, good or bad, that will have an impact on human health.

When transforming landscapes and modifying land use, we must always seek compromise: the construction of a road in a forest region of a poor country will facilitate access to markets, schools, and hospitals, but might also create new habitats for vectors of disease; the draining of swamps will eliminate the habitats of mosquitoes that transmit diseases, but will also have a negative effect on biodiversity and the hydrologic cycle; irrigation will increase food security and will thus decrease malnutrition, but will favor populations of mosquitoes and will create a new risk of schistosomiasis (or bilharziasis); the intensification of animal farming will also have a positive effect on food production, but will encourage new zoonoses; and so on. To be able to identify and choose development pathways that will maximize the positive effects for human well-being while avoiding the detrimental effects on natural ecosystems and thus on health is one of the great challenges for sustainable development.

BILL GATES OR MICROBES

Globalization has had many positive effects in the realms of the economy, culture, technology, and institutions. It has entered into every area of our daily lives, from the Internet to low-priced merchandise made in China, from our favorite television series to what we eat. However, not everyone is equal in the face of globalization, and some have benefited more from it than others. A company like Microsoft, for example, has been able to install its Windows software in a billion computers throughout the world. And yet Bill Gates is undoubtedly not the one who has profited the most from globalization: he is overtaken in this competition by microbes which, thanks to the international systems of transportation of people and merchandise, today travel around the world in a few days, and have thus gained permanent access to several billion people susceptible of being infected.

Globalization itself has not changed the natural environment, but it has enlarged it: agricultural and industrial regions in Asia or in the equatorial forests of Africa today are just outside our doors, which we can no longer hide behind. What is the impact of this on our health? New diseases take advantage of globalization to spread at dizzying speed over the entire planet. Does this threaten in the long term to cancel out the positive effects of globalization, and to relegate large multinational corporations to the ranks of minor beneficiaries of the expanded flow of goods and people among continents? Bill Gates is far from fading into the background; he devotes most of his fortune to the foundation he created with his wife, Melinda, one objective of which is to eradicate microbes responsible for diseases in poor countries. Bill Gates can already boast of great successes, including his contribution to the development of a vaccine against malaria. Perhaps the leaders of multinational corporations who become philanthropists will ultimately be the great benefactors of globalization.

The history of civilization has been marked by periods of pandemics, which are becoming increasingly frequent throughout the world. Pandemics are epidemics that affect a large proportion of the population of a continent, or even of the world. They arise either following the emergence or reemergence of a contagious pathogenic agent not recognized by the human immune system, or on the occasion of the emergence of a subtype of microbe resulting from a genetic modification of an already known microbe. The Bible mentions several epidemics, including some of the plagues on Egypt described in Exodus. Deuteronomy describes twenty years of illnesses endured by the Hittites after they imported and enslaved Egyptian prisoners of war. Those prisoners would have been carriers of microbes against which the Hittite populations, less cosmopolitan than the Egyptians, had no immunity.

The Greek historian Thucydides gives a detailed description of a fatal illness whose identity we still don't know exactly, which struck the people of Athens during the Peloponnesian War, in 430 BC. The symptoms, which Thucydides describes in vivid detail, were particularly horrible: first, a violent fever; an inflammation of the eyes, the throat, or the tongue; fetid breath and labored breathing, all followed by sneezing, diarrhea, vomiting, and violent spasms; and finally a lividity of the skin, which was covered with blisters and pustules, accompanied an unquenchable thirst. Most afflicted people died within a week, and the rare survivors lost their toes, fingers, genitals, and sight. This disease seems to have appeared first in Ethiopia, then spread throughout the Greek world, passing through Egypt and Libya. Later, in 166, the Roman army under Marcus Aurelius brought the Antonine plague (probably smallpox) back from its campaigns in the East, a disease that had 5 million victims, including Marcus Aurelius himself. During a second epidemic, Rome had 5,000 victims a day.

The Silk Road was a route for the exchange not only of textiles and spices but also of microbes: half the population in the north of the Chinese Empire died in the third and fourth centuries following the introduction of smallpox and measles from the West. In the sixth century, a new plague epidemic (the Justinian plague) struck Constantinople and Italy before traveling up to Ireland and Great Britain. Boats crossing the Mediterranean carrying wheat from Africa would have brought Justinian plague from Egypt or Ethiopia. It claimed 100 million victims in two successive waves between 541 and 767. The first wave corresponded to a cooling of the climate that lasted several years, following the eruption of an enormous amount of dust into the atmosphere from the Krakatoa volcano in Indonesia.

After a respite of a few centuries, the bubonic plague appeared in 1347 under the name of the Black Death, due to the blackening of the victims' skin from subcutaneous hemorrhaging. The disease started in China in 1330 at the latest, and slowly but surely advanced toward the West along the routes used by travelers and merchant ships. Let us note that it took some fifteen years at least to travel from China to Europe. The worried inhabitants of port cities started isolating the boats that arrived from foreign lands for a period of forty days—*quarante* in French—the origin of the term *quarantine*. The plague killed a third of the populations in northern and central Europe in the middle of the fourteenth century, and claimed more than 200 million victims in three pandemics between the fourteenth and the seventeenth centuries. France might have lost 40% of its population. These events were systematically preceded by periods of a dozen or so years during which the climate had cooled, the exact effect of which on the unleashing of the pandemic has not yet been fully understood.

The European colonization of the New World was also accompanied by its share of diseases: smallpox above all, but also measles, mumps, typhus, and flu. Historians estimate that three-fourths of the population of central Mexico died of them. The devastation caused by these epidemics was decisive in the European conquest of the Americas and the domination of the native populations: they caused many more deaths than battles and rebellions against the colonists. Microbes, moreover, often preceded the Spanish conquistadors as they advanced into the interior of the lands, so contagious were those diseases. Francisco Pizarro was believed to have intentionally unleashed such epidemics by offering blankets contaminated with smallpox to native populations. These diseases destroyed two great civilizations in Central and South America: the Aztecs and the Incas. The only disease that America might have exported to Europe at that time, in a fair exchange, was syphilis, or at least the most pathogenic strains of the *Treponema* responsible for that affliction. Crew members on Christopher Columbus's ships would have transported the infection during a return trip, in 1493. The following year, Neapolitan prostitutes infected soldiers of the French army. A few months later, those mercenaries from all over Europe brought the infection back home with them. People adopting risky sexual behavior today continue to deal with the threat of contracting syphilis.

Malaria was brought to Brazil from Africa on the boats of explorers and slave traders. Originating in India, cholera became pandemic at the beginning of the nineteenth century. Despite a vast campaign to prevent it in France, a violent explosion of cholera occurred in Paris at the end of March, 1832, with 500 new cases every day after a week, and 1,000 new cases daily a week later. The death rate was 85% among those infected, which necessitated the removal of thousands of cadavers in removal wagons that had been requisitioned for that purpose, and their quick burial in common graves. Cholera bacteria survive for a long time in water, and are often transmitted through contaminated water. No fewer than seven cholera pandemics have ravaged the world, in particular in urban zones and along major trade routes. Cholera continues to appear episodically, in places where poverty and a lack of hygiene open the door: in the slums of large cities in Latin America since 1991; in Bangladesh at times of severe flooding during monsoon rains; in Zimbabwe in 2008, following the collapse of the economy under President Robert Mugabe; and in Haiti in 2010, following the breakdown of society as a result of the 2009 earthquake.

The Spanish flu was at the center of a particularly deadly world pandemic: one-fifth of humanity was infected between 1918 and 1920, at the end of World War I. Between 50 million and 100 million people died in two years. Truly global, that pandemic didn't even spare the populations of the Arctic or the Pacific islands, which are, however, quite isolated. The rate of infection was enormous (up to 50% of all people were afflicted), and it was particularly virulent among young adults, which is very unusual. The death rate from that flu was from 2 to 20% of people infected, depending on the regions and the age groups. One of the secrets of its destructive power was the severity of associated superinfections, notably of bacterial pneumonia. A large proportion of people died of complications of the flu, which had weakened their immune system. Around 25 million people died during the first twenty-five weeks of the pandemic. By comparison, seven decades later, it took twenty-five years for acquired immunodeficiency syndrome (AIDS) to have as many victims. The Spanish flu apparently originated in China.

A macabre indicator of technological progress: in the many wars that have occurred throughout the history of humanity, it was only in World War II that the number of dead on the battlefields caused by army actions exceeded that of victims of microbes transported by enemy armies. Until the beginning of the twentieth century, waging war consisted above all of exchanging diseases rather than bullets and shells. Pathogenic agents were more effective than military strategies, especially when they infected soldiers who were already exhausted and weakened by the filth of the trenches and military encampments.

One example of such intentional biological warfare took place in 1346 during a siege of the city of Caffa (today Feodosiya) in Ukraine, during the pandemic of the bubonic plague. The Tartar army catapulted cadavers of soldiers who had died of the plague over the walls of the besieged city. The inhabitants fled the city, but they had already contracted the disease. It is likely that they had been contaminated by the fleas transported by rats, who were free to travel across enemy lines, rather than by the cadavers of plague victims.

During the Crusades, it was common for only one-tenth of an army to arrive at their destination, the rest having succumbed to typhus, plague, and other diseases picked up along the way. During World War II, Japanese troops attempted to provoke a plague epidemic in China by depositing paper bags filled with fleas infected with the *Yersinia pestis* bacterium in several cities throughout the country. This attempt quite fortunately was not successful.

More recently, the human immunodeficiency virus (HIV) responsible for AIDS advanced between 1981 and 2009, with already 60 million people infected and 25 million dead. In 2009, 33.3 million people were living with HIV. The number of deaths linked to AIDS is 1.8 million per year in sub-Saharan Africa. At present, no person infected by HIV has been cured. A bit of trivia: two of the first victims of AIDS became unintentionally famous. Following a stop in a port in West Africa, a Norwegian sailor then infected his wife, who transmitted the infection to her daughter; the members of this family had the sad privilege of being the first documented cases of the infection in Europe. And during his numerous trips abroad, a Canadian flight attendant (known as "patient zero") infected his many sexual partners in several American cities and elsewhere in the world; his nomadic sex life accelerated the spread of the disease worldwide.

The next pandemic will perhaps be that of a new strain of flu, for example if a genetic mutation of the virus were to facilitate the transmission of the infection from person to person before a vaccine is found and administered to a large portion of the population susceptible to infection. The false alerts of the H5N1 bird flu and H1N1 swine flu in 2009 may have weakened the social acceptance of a rapid public response the next time around, therefore increasing vulnerability. Or perhaps a pandemic will occur when and where no one expects it; it will then benefit from the element of surprise. Or perhaps ancient microbes trapped in Antarctica in ice formed over the last 8 million years will be brought back to life and placed back in circulation in the oceans with the melting ice as the climate warms. Humans will have no immunity against them.

Climate change could also decrease the risk of transmission of some diseases. A large number of people benefit from this every year due to the fact that the flu season occurs in the winter in latitudes with temperate climate, and not in the summer. It has recently been shown that the virus in question is more stable and has a longer life span when the air is cold and dry. The flu is transmitted most successfully when the temperature of the air is 5°C (41°F) and the atmospheric humidity is 20%, and it is less successful when the temperature rises and the air becomes more humid, stopping completely at a temperature of 30°C (86°F) and a humidity of 80%. With global warming, the flu might not be transmitted as easily, and the flu season could be shorter.

FROM MONKEY TO HUMAN

Throughout history, most pandemics have been linked to the mobility of armies and merchandise, to large concentrations of humans in cities, and to the continuous flow of goods and people between the countryside and cities. These factors have favored the rapid spread of diseases that were once confined in isolated regions. In the last few decades, two factors have considerably amplified the risk of the appearance of new pandemics: on the one hand, a rapid transformation of the natural environment which has become a source of new infectious agents; on the other, globalization, which encourages a rapid, large-scale spread of these microbes. The origin of AIDS is probably explained by the increasing number of people who visited equatorial rainforest regions that had not been inhabited up to then. Indeed, the HIV-1 virus evolved from a very similar virus affecting a subspecies of chimpanzee, the *Pan troglodytes troglodytes*, which occupy equatorial forests in the Congo basin. This virus is very widespread in wild chimpanzee communities (it affects 35% of the population), whereas it has been found only rarely in the blood of captive chimpanzees. The virus responsible for AIDS was identified for the first time in 1959, in blood taken from a man who resided in the Belgian Congo (currently the Democratic Republic of the Congo). The transmission of the virus to a human probably occurred during a hunt in the once inaccessible areas around remote forests, and by the consumption of infected monkey meat. The first "leap" of this virus from chimp to human dates to around 1900. Its circulation among human organisms would have then been facilitated by the development of the first colonial cities in the Congo, in particular Leopoldville (today Kinshasa).

Deforestation, tourism, and the extraction of tropical wood and minerals in forested zones place a growing number of people in contact with forest habitats that were once inaccessible to humans. In a survey, 60% of those living in a forest region in the Cameroon declared they had recently been in contact with the blood or other fluids of primates during an ape hunt or during the preparation of the meat. Without knowing it, these hunters often capture or kill animals that are infected or sick, because those are the ones that are unable to escape. The meat of wild animals is often sold later in local markets.

Globalization can also produce a deadly pandemic by facilitating the convergence of different serotypes of a virus in the same place. Serotypes are categories

Bill Gates or Microbes (85)

in which some viruses or bacteria are classified, by function of their reaction to a serum containing specific antibodies. They are thus distinguished by a characteristic set of antigens, or defense reactions of the organism upon encountering a foreign body or certain microbes such as viruses. They enable the differentiation of various strains belonging to the same species of microbes.

The example of dengue is enlightening. As we have seen, this disease is transmitted by *Aedes aegypti* mosquitoes. The dengue virus is a complex of four serotypes. Since the disease became known and until 1950, only one of these four serotypes of the virus was present in each of the geographic regions where dengue existed. The infection by a single serotype of the virus causes painful symptoms, with fever, pain behind the eyes, vomiting, diarrhea, and great pain in the muscles and joints, which then disappear spontaneously. Afterward, the individual develops a lifelong immunity against the serotype of the virus with which he or she was infected. During the last decades, the increased mobility of infected people and mosquitoes throughout the world, by plane for the former and by boat for the latter, has enabled the spread of the four serotypes on a global scale. The risk of infection by several serotypes of the virus has thus existed for some fifty years. Indeed, there is no cross-immunity among them: the fact of having been infected by one of the serotypes does not protect against an infection by any of the other three. On the contrary, in fact, for reasons that are still unknown today, the disease is much more serious in the event of a second infection by another serotype. As a result, there are currently devastating epidemics of dengue hemorrhagic fever.

According to the estimates of the World Health Organization (WHO), between 250,000 and 500,000 people suffer from this very severe form of the illness every year, out of the close to 50 million people infected by only one of the serotypes. Among children, hemorrhagic fever can lead to dengue shock, which causes 25,000 deaths per year. A bit less than half of humanity lives in regions where there is a risk of transmission of dengue, primarily in urbanized and semiurbanized tropical regions. The cost of globalization overwhelmingly falls to those who benefit from it the least: the most vulnerable people and children in poor tropical countries.

FROM THE PLEASURE OF TRANSCONTINENTAL TRAVEL

The emergence of a new infectious disease goes through three stages: the introduction of the infection somewhere, the establishment of the illness in that place, and the spread of the disease to other places. Currently, as we have seen, the globalization of the economy and the development of more efficient means to transport goods and people have created a powerful vehicle for the rapid spread of pathogenic agents on a global scale. Many experts agree, for example, that the spread of the bird flu is related at least as much to the international bird trade—a sometimes illegal enterprise—as to the migration of wild birds.

The contemporary period is without precedent in terms of the volume, speed, and extent of international travel. It is estimated that in rich countries, the mobility of people is a thousand times higher today than in 1800. The number of passengers-kilometers traveled by plane per year—that is, the total number of kilometers traveled by domestic and international flights per year multiplied by the number of passengers in each flight—is over 3,000 billion. The number of trips taken by plane continues to increase by 6% per year, this number having been at 9% per year since the 1950s. Air travel involves more than 5,000 airports with regular international flights. In 2006, airplanes had transported more than 2.1 billion passengers, 760 million of whom traveled internationally. Added to this is a considerable number of people who cross national borders by means of transport other than planes. The map of the spread of some diseases, AIDS for example, over U.S. cities corresponds almost exactly to the flow of air traffic. Furthermore, close and prolonged contact during long-distance flights is regularly responsible for epidemics of flu, measles, tuberculosis, and other infectious diseases transmitted through the air. Useful information for the next plane trip: the risk of contagion is generally restricted to the two rows of seats in front or in back of an infected person.

Tourism is increasingly colonizing remote and wild regions, and brings people into environments rich in infectious agents against which they have no immunity. The travel industry is experiencing a growth that is 1.3 times more rapid than that of the gross world product. Business travelers and tourists represent only a small fraction of the total number of travelers: migrants, refugees, students, seasonal workers, and pilgrims also contribute to the global spread of diseases. The pilgrimage to Mecca, for example, where more than one million Muslims from all over the world gather each year for a few days before going back home, has already contributed to the global spread of diseases such as meningococcal meningitis and poliomyelitis. These trips have a potential impact on the health of the travelers, their families upon their return, and the populations visited.

Each year Paris, Brussels, and London experience several dozen cases of "airport malaria," which affects people living near an airport. These infections are caused by mosquitoes that come on board international flights from tropical countries. Since the infected people have no immunity against the disease, the cases are often fatal.

True technological prowess has enabled the construction of bridges and tunnels that are getting longer and longer (up to 40 or 50 kilometers [25 or 31.2 miles] in length), and which also provide many species with the opportunity to cross natural barriers that were once impassable: seas, large lakes, and mountain chains.

The transportation of merchandise has also increased exponentially in the recent past, and could continue to increase in the decades to come. An ever-higher proportion of our food travels the world before landing on our plates, either because it comes from an exotic country or because it has been processed in part in places where the cost of labor is cheap. Many fruits and vegetables in our supermarkets have been picked in tropical regions just a few days earlier. This is the case for green beans from Kenya, for example. The transportation of vegetable and animal products (living or dead poultry, exotic animals) over thousands of kilometers facilitates the movement of microbes, which travel as hidden passengers in vegetation or within animal organisms. A sharp rise in meat consumption, particularly in developing countries, is an important impetus for the increase in the international transport of animals and, with them, the microbes they are carrying.

The *Aedes albopictus* mosquito came into the United States on a boat from Japan. In the laboratory, this mosquito is a competent vector for transmitting twenty-two diseases, and in natural conditions it is associated with the transmission of dengue. The Marburg virus (a cousin of the Ebola virus) owes its name to a German town where seven scientists died after working with the blood and tissue of a monkey imported from Uganda to develop a vaccine against poliomyelitis. A species of bat from sub-Saharan Africa is a reservoir of that virus. The expansion of the world trade in organs, blood, and tissue for transplantation also creates a risk for the transmission of unknown pathogenic agents.

Finally, an infatuation with wild and exotic animals such as iguanas, turtles, pythons, parrots, or felines offers a perfect opportunity for the global spread of microbes and the emergence of new diseases. The families that keep these animals in their home, their yard, or even their bedroom are often unaware that they are contributing to the threat of an emergence of zoonoses—in addition to the fact that they encourage the poaching and mistreatment of wild animals. West Nile virus probably arrived in the United States at the same time as a shipment of wild birds. Despite the health risk it presents, the global trade in exotic animals, rarely legal or ethical, is highly lucrative for unscrupulous traffickers.

Pathogenic agents are more or less capable of being effectively transported over long distances, depending on their biological demands and the way they are transmitted. A microbe that survives in the human organism, remains infectious for a long time, and is contagious through physical contact among people via corporeal excretions or sexual contact will spread easily throughout the world. This is the case, for example, of the AIDS, tuberculosis, hepatitis B, or measles viruses. By contrast, an organism whose life cycle goes through an animal host sensitive to environmental conditions (such as temperature and humidity) or which is transmitted through an arthropod vector whose range of influence is limited by the environment will have more difficulty expanding its geographic range.

A pathogenic agent will travel all the better when the "vehicle" it is living in is a good traveler. Humans have become by far the most effective travelers of the entire animal kingdom in terms of the speed and range of their travels and their tolerance for variable environmental conditions. They are, of course, helped in this by their sophisticated technological arsenal in the realms of transportation, food, air conditioning, and heating systems. For a microbe, infecting a human being is thus the most effective strategy for spreading on a global scale. The monkey and bat are much less effective and flexible travelers, and the microbes that infect them exclusively are condemned to lead a more sedentary existence.

All the same, the transport of merchandise offers another means of travel to which a mosquito or a microorganism can easily adapt. Hundreds of species of microbes (including the bacterium responsible for cholera in humans) have, for example, been found in the ballast of boats. Furthermore, the simplification of natural ecosystems and the climate changes that can homogenize places of origin and the destinations of microbial travelers also facilitate their adaptation to new lands. Indeed, the environmental shock associated with travel is then lessened. When microbes feel at home, their rate of survival at their destinations is higher. The large hotel chains apply the same principle of minimum "shock" for human business travelers through the homogenization of the environment. Wherever you travel in the world, your room is identical: the thermostat is set at 22°C (71.6°F), the bathroom is to the right of the bed, and the television is to the left. Environmental change is doing the same for microbes.

SARS, SUPER-ASIAN RAPID SURGE

The case of severe acute respiratory syndrome (SARS) illustrates how international travel by plane is an extraordinarily effective means of spreading pathogenic agents globally, thereby facilitating pandemics. In November 2002, the virus responsible for SARS appeared in the Chinese province of Guangdong among people involved in the trade of living animals, including the civet. The microbe probably mutated in one of these animal markets that are so common in China and Southeast Asia, and in which one finds reptiles, poultry, and wild or domesticated mammals cohabitating in cages piled on top of one another. It is common to see these animal stalls a few steps away from small shops with doubtful hygiene where meals are prepared for the throngs of passersby. Buyers and sellers then go back to their respective homes with a few living animals or dead animals covered in blood, attached to the racks of their motor scooters or on their children's laps. The virus responsible for SARS was first propagated locally in these places, which were very favorable to the emergence of zoonoses— in fact, ill-intentioned people couldn't have come up with a more effective means to produce new pathogenic agents than these Asian animal markets.

On February 21, 2003, while he was suffering from the first symptoms of the disease, a Chinese doctor from Guangdong spent a night on the ninth floor of the Metropole Hotel in Hong Kong. He died a few days later. Sixteen of the hotel guests who had spent the night in rooms on the same floor were infected. Since the incubation period of the virus is 2–7 days, those people led normal lives for the first few days after their infection, contaminating others along the way. Some of them took planes to go home to Toronto, Singapore, Vietnam, Ireland, the United States, Germany . . . the infection then caused new cases among the passengers of those planes and in the immediate entourage of the infected people, wherever they stayed after their contamination. The pandemic was launched. Between March and May 2003, or for just three months, more than 8,000 probable cases of SARS were reported, with more than 700 deaths, in 26 countries spread over 5 continents. Each infected person engendered on average 2.7 new infections. More than 90% of the cases were cured after 5 or 6 days of being diagnosed.

The disease was identified for the first time by doctors in the French hospital of Hanoi, in Vietnam. A worldwide collaboration coordinated by the WHO isolated the agent that causes SARS (a new virus from the Coronaviridae family) on April 16, 2003, less than two months after it appeared in Hong Kong. Getting the word out, and, when cases were suspected, the examination and potential isolation of international travelers originating from the contaminated regions, seem to have been crucial measures taken to control the epidemic. The epidemic seems to have been responsible for an approximately 2% decrease of the gross domestic product of east Asian countries, and that economic impact no doubt had human consequences that were even more serious than the 700 deaths directly caused by the virus.

The SARS virus was transmitted mainly through direct contact between individuals, by the ejection of infected microdroplets originating in the lungs. A transmission through the environment, through sewer water in particular, cannot be ruled out. The horseshoe bat (*Rhinolophidae*) from China is probably at the origin of the epidemic. This species is a healthy carrier (a carrier of the virus without suffering any health effects) of a virus almost identical to the coronavirus of SARS. In other words, this bat is a reservoir of the original virus. The existence of such a reservoir maintains the virus in the environment, which enables its unexpected reappearance in humans.

Remember the Black Death that appeared in Europe in 1347 after taking around fifteen years at least to get there from China? Six hundred fifty years later, the SARS virus made the same trip in a few days, which corresponds to an acceleration of the speed of the spread of diseases by a factor of 1,000. This is one of the direct consequences of globalization.

THE GLOBALIZATION OF MICROBES

In 1997, Joshua Lederberg, 1958 Nobel Laureate in Medicine, estimated that in our continual struggle against diseases, the massive displacements of people associated with globalization have clearly tipped the scale in favor of microbes. Globalization has transformed humans into a very different species than they were a century ago. Have we thus gone from *Homo sapiens sapiens* to *Homo sapiens mundus*? And has this latter become intrinsically more vulnerable to the threats of pandemics? Granted, today we have highly sophisticated technology to defend ourselves against the constant harassment of pathogenic agents: vaccines, antibiotics, diagnostic tools, and the means to exchange medical information in real time. But the circulation of microbes on a global scale, in a few days, has fundamentally changed the balance of power.

What is new today in terms of threats to our health comes not from emerging diseases and pandemics—we have seen in this and the preceding chapters that they have affected all the great civilizations in the past—but indeed from the increased opportunities for emergence that the current mode of development offers them, and the acceleration of the speed at which these diseases are spread throughout the world. Up to now, the globalization of public health systems, of which the WHO is the most visible face but certainly not the only one, has enabled us to respond adequately to the threat. The maintenance of our well-being will occur at the cost of a sustained investment in the means of defense against emerging diseases. We must act so that the globalization of public health is at least as rapid as that of microbes. This is one of the conditions that in the long term will ensure that the expansion of our everyday environment thanks to globalization can lead *in fine* to a positive outcome for the well-being of humanity.

PARADOXICAL CITIES

Urbanization is the most radical form of change to a natural environment: nature is obliterated, the ground is covered with concrete, water is channeled, the night is lit up. A universe of artifacts is substituted for natural ecosystems. Nature's retreat is complete: the city is anti-nature. All the same, from a wider perspective, urbanization is the most efficient and the most ecological means of occupying the planet: it concentrates more than half of humanity onto less than 1% of inhabitable land. In fact, all the cities in the world occupy only a small area, even if their ecological footprint extends widely beyond their borders. This geographic concentration of human activity creates large economies of scale in the production of energy and transportation, and the provision of services. This is the first paradox of cities: the spatial concentration of populations destroys nature locally, but increases the ecological effectiveness of our economies.

Cities have always been places where people can flourish, thanks to dense and multicultural social networks of innovation, political participation, cultural change; places of more supple social norms; and access to education, information, and services. This is what makes them so attractive. And yet cities have taken and still take a heavy toll on health and well-being. Today, poverty, violence, a lack of safety, and diseases are part and parcel of urban living, particularly large cities in developing countries. This is the second paradox of cities: the power of attraction of the city lights goes hand in hand with an elevated risk in terms of individual well-being. In this chapter, I will look at the impact of urbanization on health.

Since 2008, a bit more than half the world's population lives in cities. In 1800, only 3% of the population was urban, and 14% in 1900. Today, some twenty cities have more than 10 million inhabitants. By 2030, most global demographic growth will occur in cities in developing countries, while the rural population will decrease slightly. By that date, the urban population will be 6.4 billion inhabitants (or 70% of the world's population).

As has always been the case throughout history, urbanization is the engine for economic development. In 2008, cities accounted for 75% of the world's energy consumption, but also 80% of greenhouse gas emissions. The solution to the great challenges that humanity must face in the coming decades—stabilizing the world's population, eradicating poverty, deploying sustainable energy systems, diminishing the ecological footprint of human activity—necessarily goes through cities, which are the focal points of globalization.

In the last few decades, rich countries have experienced a wave of periurbanization among the wealthier social classes looking for a better way of life in the countryside close to cities. However, on a global scale, the absence of massive migrations from cities to rural areas suggests that the potential benefits of urbanization outweigh its disadvantages. And yet, today as in the past, cities in poor countries are the home of various infections, as they foster deplorable living conditions. A third of the residents of cities in developing countries live in slums. In the large cities of sub-Saharan Africa, the number can reach as high as 70% of the population; they live in conditions of extreme poverty, without access to running water or to health systems: an average of three toilets and one shower for 250 families is common in these slums. The use of "flying toilets" is very widespread: people defecate in little plastic bags that are then thrown onto a pile along the roadside. Infant mortality linked to contaminated water is very high. The prevalence of diseases caused by a lack of hygiene as well as acquired immunodeficiency syndrome (AIDS), often combined with tuberculosis, means that in certain slums (in Nairobi, for example), close to a quarter of the children do not survive beyond the age of five.

In rich countries, problems affecting socially marginalized people (alcoholism, drugs, crime), as well as depression and illnesses linked to stress and isolation, including sexually transmitted diseases, are more frequent in cities than in the country, either because urban living conditions foster such problems, or because the people that endure them for the most part choose to live in the city.

Another paradox of cities: although many health problems are concentrated in them, in poor countries mortality is often lower in cities (as much as 20% less) than in rural areas. There are two reasons for this: on the one hand, in these countries city dwellers have better access to quality health services, of which there are many more in cities than in remote rural areas; on the other, there are enormous social inequalities in cities in developing countries, where the poor of the slums, whose life expectancy scarcely exceeds 30 years, live alongside a privileged social class that benefits from health care on a par with that in the West. This wealthy fraction of the urban population decreases the average statistics for the mortality rate in cities. Unlike in European cities up until the nineteenth century, the birthrate in most cities in poor countries is higher than the death rate. The resulting growth of the urban population by itself explains 60% of the population increase of cities in developing countries. The remaining 40% can be attributed not only to migrations from rural areas to the cities but also to the integration of rural populations into the urban sphere as cities expand.

THE CESSPOOL OF THE FIRST CITIES

The formation of the first cities, around 5,000 years ago, was associated with a rapid growth of infectious diseases, such as tuberculosis, leprosy, and diarrhea. In fact, an increase in the number of people likely to be infected, as well as the density of a population, increases the risks that a pathogenic agent will survive and multiply by being transmitted from one host to another; such population increases thus favor the propagation and establishment of diseases. The risk of reinfection also increases, as do the quantity and diversity of pathogenic agents that an individual can encounter during his or her life. In the case of factors leading to diarrhea or cholera, for example, the infection is transmitted by oral contact with water, food, objects, or hands that have been contaminated by fecal matter. Whereas nomadic communities left their contaminated camps at regular intervals, thus leaving environments favorable to the development of pathogenic agents behind them, by contrast, urban societies live permanently in immediate proximity with their excrement, which increases the "fecal peril." An accumulation of trash also attracts rodents and their parasites, which can transmit bubonic plague, the human form of typhus, and several types of hemorrhagic fever into human habitations.

The first human populations, which lived in small isolated groups, probably experienced only a limited number of infectious diseases—chronic illnesses with minor symptoms, rarely fatal. In fact, in terms of biological evolution, a pathogenic agent causing a rapid death of its hosts would not be very "efficient," because it would condemn itself to extinction, for lack of new hosts to infect to ensure its survival and proliferation. Epidemics of serious diseases could only be short-lived and limited to a few communities, and would have disappeared by themselves once all the potential victims had been stricken. As an analogy, a fire on a little isolated patch of grassland cannot spread, and thus spontaneously extinguishes for lack of new fuel. However, when a very large reservoir of potential hosts in perpetual contact with one another became available and was constantly renewed—which was the case with the appearance of the first cities—infectious diseases with much more serious symptoms began to settle in. Then, from the microbe's point of view, it didn't matter if a host died, since there was a constant flow of new hosts. Another condition for the success of these diseases is that they were very contagious—in fact, they had to be transmitted quickly to other individuals before the infected people died or acquired a permanent or temporary immunity to them.

At the beginning of urbanization, the death rate was higher in cities than in the countryside, and a constant flow of rural immigrants was necessary to avoid a demographic collapse of cities. This situation continued through the end of the nineteenth century when, in England for example, infant mortality in the cities was more than twice that of the countryside. During the Black Death, which struck Europe in the fourteenth century (from 1347 to 1350), 25 to 30% of the population died, but in the cities, the death rate reached 50 to 70% of the population. One medieval prayer implored: "Save us, O Lord, from famine, war, and the plague!" Because when one of those three ills struck, the two others were not far behind. Preindustrial cities in the seventeenth century, such as London, Paris, or Amsterdam, were great hotbeds of typhus, tuberculosis, and cholera. Public hygiene was almost nonexistent: they were without systems providing clean drinking water, and the sewers were out in the open; human and animal excrement were thrown into the canals and rivers, which were also sources of drinking water.

In London in the mid-nineteenth century, the Thames was a vast cesspool that served as a sewer for 3 million people, fermenting in a fetid stew, to the point that the river was nicknamed "the Great Stink." The smell was so unbearable that in 1858, sessions in the House of Commons, on the banks of the Thames, had to be interrupted. Paris must not have been much cleaner. In these cities, several families lived in the same house and sometimes in the same room. With the deplorable hygiene conditions, poor ventilation, and people living alongside domesticated animals, rats and other vermin contributed in forming an ideal milieu for the propagation of infectious diseases. The immune system of the inhabitants was further weakened by malnutrition. According to Friedrich Engels, the German socialist philosopher and great friend of Karl Marx, the threat of infectious diseases in the poor neighborhoods of industrialized cities fostered the growth of communism in the nineteenth century.

In the seventeenth century, John Graunt, a rich textile merchant from London, for twenty-two years devoted his leisure time to keeping scrupulously detailed statistics on public health in his city. Every week, London published death records that notably specified the number of deaths per parish, gender, and cause, in order to detect plague epidemics. Considered today to be one of the first demographers, Graunt came up with the idea of classifying these data in the form of tables to detect and measure arithmetic regularities. He thus hoped to reveal scientific laws similar to those extolled by the physicists and astronomers of his time. His observations show that 30% of the population died before the age of 5 years, and half before the age of 15; scarcely more than 10% of Londoners lived beyond their fortieth birthday, and only 7% to more advanced ages. John Graunt was one of those privileged few, since he lived to be 54. Diseases such as multiple infections leading to diarrhea, measles, and whooping cough killed 10 to 20% of infants before they reached their first birthday. On average, close to 70% of deaths were caused by infectious diseases, 6% of which were from leprosy. It was only toward the mid-nineteenth century that public hygiene considerably improved in large Western cities. The invention of flush toilets with an accompanying sewer system has no doubt contributed to the well-being of humanity more than any other technology.

The positive link between the environment and human health, safety, and welfare started to be perceived by thinkers and activists during the nineteenth century. "Positive environmentalism" became a driving force in urban reform and environmental preservation. The response to the "evils" of the industrial city took on a variety of forms, from the religious to the scientific, and yielded decades of sustained action in public policy and regulation to improve the everyday environment, particularly in cities. After centuries of neglect of environmental conditions, it was finally recognized that domesticated nature and well-planned cities do contribute to physical, mental, and spiritual well-being.

Today, sanitary conditions in the slums of developing countries are scarcely better than those that existed in the seventeenth and eighteenth centuries in European cities. A bit less than half the population still does not have modern toilets. This situation forces 2.6 billion people to live in unhealthy and inhuman conditions. Some fifty diseases are transmitted through human excrement, which is hardly surprising when we know that 1 gram (.03 pounds) of fecal matter contains up to 10 million germs. Fostered by a lack of hygiene, diarrhea each year kills 1.6 million children throughout the world. In the face of this global health crisis, the United Nations declared 2008 the International Year of Sanitation, which was acknowledged mostly with indifference.

During the last century, the global average life expectancy doubled, and the health of the human population improved overall. This major progress is very largely the result of a decrease in infectious diseases, due to improvements in food safety, nutrition, hygiene, food refrigeration, education, and a host of medical advances (such as vaccination and medicines, including antibiotics). As the public health specialist Tony McMichael shows in his book, *Human Frontiers, Environments and Disease*, technological and social progress, as well as the public health infrastructures underlying an improvement in public health, is strongly connected to urbanization, industrialization, and the consumption of material goods.[1]

The urban lifestyle in industrialized societies is, however, associated with a series of new health threats that are no longer caused by infectious diseases. The principal causes of morbidity and early death in industrialized regions of the world today are cardiovascular disease, cancer, diabetes (notably linked to obesity), and depression and other mental illnesses. Industrial accidents and driving accidents are also a new source of death and injury. In the second half of the twentieth century, a new generation of environmental changes became apparent, in particular global warming, changes that have possible impacts on human health. Could progress ultimately be only the replacement of one type of health problem (infectious diseases) with another (the effects of industrial pollution and the urban, sedentary lifestyle)? If the answer is yes, the global outlook must be quite positive, because otherwise life expectancy would not have increased so dramatically.

And yet the poorest populations of the planet have the sad privilege not of substituting one evil for another but of accumulating all types of health problems: the marginal populations mainly of Asia and sub-Saharan Africa still suffer from malnutrition and infectious diseases (including AIDS, malaria, tuberculosis), and in addition must deal with the atmospheric and industrial pollution in their cities. Let's recall the disaster in Bhopal, India, in 1984, the 3,800 dead and 360,000 victims who became incapacitated and helpless to varying degrees, poisoned by a highly toxic leak of cyanhydric acid from a pesticide factory. Driving accidents every year kill more than 1 million people in developing countries, a figure that can only increase in the short term with the increasing number of

1. T. McMichael, *Human Frontiers, Environments and Disease: Past Patterns, Uncertain Futures* (Cambridge: Cambridge University Press, 2001).

cars being put in circulation, until drastic driving policies for road safety are adopted. Populations in poor countries are also the most vulnerable in the face of climate change.

Another example: in poor countries, rural families suffer above all from air pollution caused by stoves inside their houses used for cooking and heating. The smoke from these little stoves fed with wood, charcoal, or agricultural waste is in general badly ventilated and, over the years, causes serious respiratory diseases. It in fact contains many harmful pollutants, including fine particles and carbon monoxide. In rich countries as well as in the wealthy neighborhoods of poor countries, houses are equipped with smokestacks and have electrical systems, which remove that source of low indoor-air quality. By contrast, the poor neighborhoods of cities in poor countries lack electricity and ventilation systems adapted for houses, so that urban pollution of the outside air that is associated with industrialization is added to the stove smoke inside homes. Here, too, the old pollution is not replaced by the new; the new is added to the old.

Let's now look at two typically urban health risks: air pollution and heat waves.

UNBREATHABLE AIR

Air pollution has negative effects on human health. It is caused by gases or particles whose chemical properties irritate the respiratory system, particularly that of asthmatics: ozone; nitrogen dioxide; sulfur dioxide; fine suspended particles (PM, for particulate matter), noted as PM10 when their diameter is less than 10 micrometers, or 10–6meters, and PM2.5 when it is less than 2.5 micrometers, some of which are carcinogenic; benzene, also carcinogenic; carbon monoxide, deadly in high doses. The World Health Organization (WHO) estimates that 3 million premature deaths per year (or 5% of deaths per year throughout the world) can be attributed to the effects of atmospheric pollution. There are about 400,000 such deaths per year in the European Union, where long-term exposure to PM2.5 reduces life expectancy by 8.6 months. In Europe at the beginning of the 2000s, deaths attributable to air pollution were six times higher than deaths caused by driving accidents. Air pollution increases mortality and morbidity linked to respiratory and cardiovascular diseases, causes acute asthma attacks, increases the risk of lung cancer, and is a source of fatigue, headaches, respiratory infections, and eye irritation.

The problems of air pollution appeared on the front pages of newspapers in the period leading up to the 2008 Olympic Games in Peking. It was only after that city put its economy and road traffic essentially on hold for several months that the amount of pollution was low enough for sporting events to be held there. An atmospheric phenomenon that was unusual for the inhabitants of Peking was then observed: the sun was no longer permanently hidden behind a thick brown cloud of pollution, and the sky was even occasionally blue.

China has 670 cities, 89 of which have a population of more than 1 million people. In 2005, those cities contained 40% of the Chinese population and contributed 65% of the country's gross national product. In 2020, 60% of Chinese will live in cities, which will make that wave of urbanization the greatest migration in the history of humanity. Do the names Shenzhen, Dongguan, Tianjin, or Chongqing mean anything to you? Probably not, and yet in 2007 those modern and industrial cities all had between 5 million and 10 million inhabitants. And the large Chinese cities are among the most polluted in the world. A survey done at the end of 2008 in 320 Chinese cities revealed that the air of two urban zones out of five was polluted, and was even considered dangerous. The problem had worsened by 2010.

China still uses coal for 70% of its energy consumption. Emissions associated with the burning of this coal, including sulfur dioxide (SO_2), responsible for acid rain, are a significant source of air pollution. One of the objectives of China's tenth five-year plan, which debuted in 2001, was to reduce the sulfur dioxide in the air by 10%. Five years later, that pollution had increased by 27%. Furthermore, given the explosion of construction in Chinese urban regions, heavy industry (cement, steel, and aluminum production) has also become very polluting.

Furthermore, again in China, the number of cars in cities has increased by 10% per year. In Peking, concentrations of very fine particles in the air ($PM2.5$), which are extremely harmful to health, have reached levels that are ten times higher than the WHO norms. Nitrogen dioxide (NO_2) and ozone (O_3) contribute to the main problems of air pollution. Concentrations of nitrates (NO_3^-) and of ammonia are also very high. In that country, air pollution is the cause of 750,000 premature deaths per year, essentially the results of respiratory and heart diseases—a figure that corresponds to the elimination of the population of an average city each year!

Faced with this disaster, China has launched a half-dozen experiments in the construction of eco-cities, such as Dongtan, near Shanghai. These will be new urban centers whose ecological footprints will be half those of normal cities, and they will recycle 90% of their waste. All energy needs will be met through renewable sources. Only electric and hydrogen cars will be allowed. The air should be more breathable.

In North American and European cities, air pollution first increased with industrialization, then during the last thirty years quickly decreased, in particular sulfur dioxide (SO_2) and carbon monoxide (CO) pollution. This decrease is owed to the adoption of less polluting technologies; stricter, vigorously implemented environmental legislation; and an increased willingness of consumers to pay for clean technologies. Furthermore, some of the polluting industries in rich countries have been moved to emerging countries, such as China. Exporting one's pollution means importing well-being. This begs the question: where will China export its pollution to improve the quality of the air in its cities? To Africa, probably, within the next few decades.

DYING OF HEAT IN CITIES

The urban environment also has an impact on health through what is called the urban heat island: throughout the year, temperatures in cities are on average from 0.5° to 1.2°C (32.9° to 34.1°F) higher than in neighboring rural areas. This difference in temperature can reach 5° to 12°C (41° to 53.6°F) in a given period in some large cities. The urban heat island results from a decrease in plant cover, since plants have a cooling effect on the earth's surface through transpiration, and from a decrease in wind speed due to obstacles such as buildings. Furthermore, darker built surfaces absorb more heat in daytime than lighter natural surfaces. Everyone has experienced the heat emitted by asphalt roads at the end of a hot, sunny day. Much less relief is brought by a temperature decrease at night in a city, since the heat absorbed during the day is then reemitted. Finally, the production of heat through human activity, notably by vehicles and air-conditioning systems in buildings, also contributes to this warming.

The human body has an adapted comfort zone that corresponds to maximum temperatures between 19° and 29°C (66.2° and 84.2°F), depending on the humidity of the air and to how acclimated a population is to the normal temperatures of its environment. The types of infrastructures in cities, such as the presence of air-conditioning systems and the architectural characteristics of houses, also influence that comfort zone. Once a temperature threshold has been passed, there are an increasing number of deaths, caused by cardiovascular and respiratory diseases, dehydration, and hyperthermia.

According to the World Meteorological Organization, a heat wave is a prolonged period of abnormally and uncomfortably hot and usually humid weather over a large territory, generally lasting several days to a few weeks. The precise temperature thresholds and duration vary from one country to another, but in general a heat wave implies a maximum temperature of over 30°C (86°F) for forty-eight hours or more. In California for example, "heat storms" occur when the temperature reaches 38°C (100°F) for three or more consecutive days over a wide area. Heat waves often act in synergy with air pollution, having an increased impact on health. For example, the concentration of ozone at low altitudes increases with the temperature, and mortality during heat waves is higher during days of high concentrations of PM_{10}. During the great heat waves of 1976 and 1995 in Great Britain, the death rate in cities was higher by half than what was observed over the entire country, after taking into account age differences in the populations of cities and rural areas (urban populations in fact have a slightly higher proportion of the elderly, who are more sensitive to abnormally high temperatures, than do rural populations).

In general, victims have preexisting health problems, and so the smaller heat waves have a "harvesting effect," an expression that suggests a short-term acceleration of mortality. In those cases, mortality decreases during the weeks that follow the heat wave, compensating for the deaths that occurred prematurely. Elderly people are the most affected, and this has a great impact on regions of the world that have an aging population, such as Europe. The heat waves that are most threatening to health are those that appear early in the season, last for several consecutive days, and are associated with high temperatures at night, thereby depriving bodies of a daily reprieve.

The effect of this urban heat island is exacerbated by global climate change, which according to the Intergovernmental Panel on Climate Change is associated with a greater variability in the climate and thus to an increased frequency of extreme meteorological events. Let's note that human mortality increases more quickly with abnormally high temperatures than with exceptionally low temperatures, to which it is easier to adapt one's behavior, clothing, and infrastructures. The decrease in mortality linked to global warming during the winter is thus unlikely to compensate for the increase in mortality that this warming causes in the summer. Furthermore, in regions with warm climates, the incidence of some infectious diseases, cholera and salmonella for example, increases greatly with rising ambient temperatures.

In western Europe, the frequency of heat waves during the summer has almost tripled since 1880, and their duration has doubled. In Europe during the second half of the twenty-first century, these heat waves will become more intense, more frequent, and longer lasting. The heat wave that struck there during the first two weeks of August 2003 caused 40,000 deaths, including 15,000 in France. It has been shown that there was no "harvesting effect": mortality remained high in the months that followed the heat wave. It was the hottest summer observed in the last five hundred years, with an average temperature 3.5°C (38.3°F) higher than the average over the long term, and temperature spikes over

35°C (95°F). Most deaths occurred in urban areas, such as Paris. They involved for the most part the low-income elderly who had previous medical conditions and lived alone in small apartments on upper floors of their buildings. There were more women than men who died. The number of related deaths was larger in the north of the country (150% in Paris) than in the south (25% in Marseille). By contrast, the heat wave of July 2006, which was certainly less severe, caused only 112 additional deaths in France. This difference is explained by a series of preventive measures taken by the public authorities within the framework of the heat wave plan launched in 2004, which attests to a society's great ability to adapt in the face of new health risks. Yet, the 2010 Russian heat wave, which affected the entire Northern Hemisphere, caused 56,000 deaths, including those associated with forest fires and drought. By late July and early August, numerous cities in western Russia, Ukraine, Belorussia, and the Baltic states experienced record-breaking temperatures near 40°C (104°F), more than 10°C (50°F) warmer than normal temperatures at this time of year. That heat wave was an abrupt and extreme event for which these countries were unprepared.

The relationships between environmental changes and health should thus be analyzed in terms of vulnerability: what risks of harmful effects does an external disturbance pose to a person or a community? Vulnerability, or the lack thereof, depends on factors that enable people or groups of people to anticipate, face, resist, and recover after a disturbance. Taking vulnerability into account requires an integration of economic, social, and political strategies of adaptation, and not simply considering the physical, biological, or chemical changes in the natural environment. In other words, the impact of a heat wave depends not just on the spike in temperatures compared with what is normal but also on the socioeconomic conditions of the people who endure it. In fact, vulnerability in the face of a heat wave depends on three groups of factors. It varies by means of one's exposure to the risk of excessive heat, which notably depends on one's place of residence, whether it is located at some altitude, in the shade of a forest, or along a water source, or, on the contrary, in a densely populated urban zone or in a valley where air circulation is poor. It also depends on a person's sensitivity to that risk, which is lower for young people in good health than for elderly people or those whose immune system is already weakened. Finally, it also depends on a person's or a society's ability to respond: whereas people who have a certain level of income can diminish the physiological stress associated with high temperatures by installing an air-conditioning system and by going to the country or to the pool on the hottest days, isolated people without means of transportation have no way to escape the stifling heat.

Other noninfectious diseases are also associated more or less directly with global warming. For example, kidney stones are more frequent in warmer regions, and cardiovascular diseases are linked to an increase of ozone concentrations at low altitude in periods of high heat.

HEALTHY PLACES

The urban lifestyle produces a population that is increasingly physically inactive, and whose food is rich in meat and sugar. The urban transportation system, the proximity between living spaces, shopping, and work, and the type of leisure the city dwellers practice do not favor physical activity. This causes a decline in public health in cities throughout the world. A typical illness in cities is type 2 diabetes (also known as adult onset diabetes or fat diabetes), which increases the risk of cardiovascular problems. It is caused by the introduction of a high amount of sugar into the blood (hyperglycemia) over a long period. It typically strikes the adult either suffering from obesity or who is overweight, who is not physically active, and whose diet is lacking in fiber and is rich in saturated fats, sugars, and animal protein. This description corresponds to the Western diet, which is quickly becoming global. Type 2 diabetes is almost nonexistent among physically active people. The increase in obesity among children has led to the fear that this illness will increase dramatically in the near future. The WHO estimates that it already affects 180 million people throughout the world, and that this figure could more than double by 2030. It is not an exaggeration to speak of an epidemic in that case.

How can we conceive, construct, and maintain healthy living spaces: homes, parks, and cities that promote a healthy lifestyle? Urban planners and public health specialists must collaborate to deal with the health problems linked to the typical lifestyle in modern cities. Already Frederic Law Olmsted, the famous nineteenth-century American landscape architect—he designed, among other places, New York's Central Park and numerous university campuses—considered human health central to his work. The materials used to construct a building, its ventilation system, and the humidity of its air determine the concentration of interior pollutants. The presence of an attractive interior stairway, well-lighted and located at the center of an office building, promotes physical activity. Natural light and the proximity of plants have a calming influence. Well-conceived urban parks—accessible, attractive, well-lighted, maintained, safe, and equipped with benches, drinking water, and toilets—promote a contact with nature and social relationships among inhabitants of the same neighborhood. It has been shown that the absence of a dense social network and the stress linked notably to a feeling of insecurity predispose people to health problems. The proximity of stores selling fresh food, sports infrastructures, and health centers promotes the adoption of preventive healthy behavior.

Wide sidewalks and bike paths protected from automobile traffic encourage city dwellers to leave their car in the garage, which decreases urban pollution and encourages physical exercise. These activities increase psychological well-being and reduce the risks of cardiovascular disease, type 2 diabetes, obesity, and depression. Furthermore, the promotion of public transportation systems increases the independence of elderly people, children, and people without cars. Finally, revitalizing city centers, those that have been abandoned in many large American and European cities, avoids the emergence of dangerous neighborhoods concentrating poverty and criminality.

The concept of the compact city has influenced land development policies for several years. The goal is to concentrate the development of cities spatially by placing new construction within them or on their edges, constructing high-density tall buildings. The objective of this type of planning was based initially on economic and social considerations, then on fostering sustainable development: it was hoped that a densification of activities would reduce car trips (and thus decrease CO_2 emissions), would encourage walking and bicycling, and would spare land for nature in neighboring rural areas. In reality, experience shows that the use of cars decreases only moderately in compact cities, unless a mixture of residential space, services, and jobs is encouraged, and an effective system of public transportation is developed. The choice of using one's car or not remains linked to income, level of education, age, type of job, and cost of fuel.

A tall high-density dwelling does not lend itself to the installation of solar panels, because the surface area of roof per inhabitant is small and the shadows from neighboring buildings diminish the panels' effectiveness. And globally, CO_2 emissions linked to energy consumption in buildings have increased by 3% per year in the last few years. In addition, a higher density in buildings causes an increase in the sound and air pollution associated with the roads that are frequently near the buildings. The increase in this sort of pollution in cities conceived to be more sustainable constitutes the paradox of compact cities. A concern with resolving an environmental problem sometimes creates others, which can have even more harmful effects on health.

Finally, the concept of a compact city pushes green spaces and parks outside the urban milieu, which engenders other negative effects on psychic and physical health, those associated with the absence of contact with nature and the difficulty of practicing sports activities outdoors. Several studies reported a lower prevalence of a number of disease clusters, in particular those related to mental health, in living environments with green space within a 1- to 3-kilometer (.62- to 1.8-mile) radius. Furthermore, the elimination of large green spaces in the city removes the alleviating effect plants can have on the urban heat island. Moreover, in a city, green spaces slightly improve the quality of the air, because plants absorb certain atmospheric pollutants, even if they have only a negligible effect on concentrations of fine particles and nitrogen dioxide.

A NEW ALLIANCE

Future urbanization, which is inevitable, will have many potential benefits for well-being, but will also be the source of new health risks. Urban growth must be conceived to find the best possible compromise between socioeconomic and environmental benefits on the one hand, and an improvement in the well-being and health of city dwellers on the other. Managing the impact of urbanization on our quality of life requires an attention to factors other than the density of buildings and distances to be traveled. A more holistic approach must be adopted, one that considers well-being, physical activity, pollution, place of residence, modes of travel, and urban planning. A close collaboration between health specialists on the one hand and architects, entrepreneurs, real estate promoters, and those responsible for land development and public transportation on the other is the best means to ensure that the built environment maximizes benefits to well-being. Adding policies that bring the rural milieu into the mix will in addition enable the rate of urban growth to be controlled.

ENVIRONMENTAL CONFLICTS

This and the following chapter will explore the impact of environmental changes on security, another essential component of a happy life. Happiness includes the assurance that we will continue to enjoy what is most important for our survival and well-being. Feeling sheltered from danger is the basis for a feeling of security. That feeling can be threatened by elements in our everyday environment, by natural disasters and, more dramatically, by violent conflicts. In this chapter, I will examine the following question: do environmental changes threaten to engender new conflicts that will in turn threaten the security and thus the happiness of us all? These would be conflicts between states and within states; and security is understood in a rather literal sense as protection from danger and the peace of mind that results from an absence of any danger to fear, and not in the larger sense of a liberation from material need, freedom, and an ability to adapt in the face of changes to one's environment.

We will discover in the following pages that although they have direct and serious impacts on health, changes in the natural environment do not contribute significantly to the most extreme forms of insecurity. The influence of the environment on the threats of armed conflicts is often exaggerated in the rhetoric of people as respectable as the secretary-general of the United Nations (UN). Although their intentions are praiseworthy, concerned as they are with promoting sustainable development, such simplifications and occasional exaggerations are harmful to the cause they are defending: they harm the credibility of the environmental movement and thus unintentionally serve the interests of groups with dubious geopolitical motives.

At the end of the eighteenth century, the Reverend Thomas Malthus predicted that famines, illnesses, and wars would arise alongside an increase in the human population, which was growing more rapidly than that of food production.[1] Although famines, epidemics, and conflicts did indeed occur, history proved Malthus wrong as to their causes, because technological innovations, education, improved institutions, and an adaptation of human behavior have enabled a considerable increase in the resources available for humans in the face of economic development. For example, between 1960 and 2000 world food production increased more rapidly than the global population—whose rate of growth, however, experienced a historic spike in the 1960s.

In spite of that, Malthus's reasoning has been picked up today in a slightly modified form. Globally, variations in the climate have had an impact on agricultural production. In densely populated regions, a lack of food has led to an increase in prices, which throughout history has caused armed conflicts, famines, and epidemics. This mechanism might explain that over the period 1400–1900, a synchronization between the rise in average temperature, agricultural production, the price of wheat, the human population, and the number of wars has been observed.[2] The end of the seventeenth century and the beginning of the nineteenth century, two periods with below-average temperatures unfavorable to agriculture, coincided with great political upheavals, in both China and Europe, whereas the eighteenth century, with milder temperatures, experienced relative calm in those two regions. And so, according to the argument, the decrease in available resources as a result of climate change had consequently led to an increase in armed conflicts.

For anyone adequately versed in the complexity of social and political phenomena, and aware of the obscure and complicated twists and turns of history whose dynamics are formed by shifting alliances, quests for power, ethnic and religious conflicts, and romantic intrigues, this macro-historical view can only appear simplistic. In addition, five hundred years is a very short amount of time to measure the correlation between cycles whose periodicity, moreover, is a hundred years. From 1400 to 1900, only a single warm period was recorded—the

1. See T. Malthus, *An Essay on the Principle of Population*, 1st ed. (London: J. Johnson, 1798).

2. See D. D. Zhang, P. Brecke, H. F. Lee, Y.-Q. He, and J. Zhang, "Global Climate Change, War, and Population Decline in Recent Human History," *Proceedings of the National Academy of Sciences* 104, no. 49 (2007): 19214–19.

eighteenth century (whereas the warm period called the Medieval Climatic Optimum occurred from the tenth to the fourteenth century), between two cold phases, the sixteenth and seventeenth centuries, then the beginning of the nineteenth century, which earned the period from 1550 to 1860 the name Little Ice Age in the Northern Hemisphere. Over such a limited time span, there is a good chance that any correlation is spurious. Today, the question of the connection between climate changes—or, more broadly, changes in the natural environment—and political instability leading to violent conflicts continues to interest historians, politicos, and the military responsible for strategic analyses.

ENVIRONMENTAL SECURITY

Presided over by Norwegian Gro Harlem Brundtland, the UN World Commission on the Environment and Development is famous for having introduced in 1987 the concept of sustainable development, as seen in Brundtland's book *Our Common Future*. This work contains the following assertion: "Nations have often fought to assert or resist control over war materials, energy supplies, land, river basins, sea passages and other key environmental resources."[3] This statement sums up well a widespread view on the origin of armed conflicts: the struggle for access to and control of natural resources would be one of the profound causes of tensions and conflicts between nation-states. Inspired by the idea that environmental changes threaten peace, environmental security has become a new field of scientific investigation. Military strategists are convinced that climate change could represent a threat to international security, one that could be as dangerous and more difficult to control than the arms race during the cold war, nuclear proliferation, or the terrorist threat by extremist Islamic groups. When it held the presidency of the UN Security Council in April 2007, the United Kingdom added the central role of climate change in matters of security to the agenda of issues to be discussed.

Environmental changes are perceived as providing a fertile ground for extremism and terrorism, and as an amplifier of threats already looming in fragile regions. Environmental degradation weakens agricultural production and increases unemployment, which leads to a weakening of rural economies, the erosion of traditional social networks, and the migration or displacement of a higher proportion of the population. And so ethnic conflicts would arise, alongside independent movements and rebellions, genocides, guerilla warfare, the

3. G. Bruntland, ed., *Our Common Future: The World Commission on Environment and Development* (Oxford: Oxford University Press, 1987), p. 290.

trafficking of natural resources extracted illegally, terrorism around mineral mines, and the migration of the young unemployed to the slums of large cities where violent gangs rule. Governments that are already fragile would then become more destabilized, which would open the door to more authoritarian regimes and to a greater political instability in relations among countries. Within nations, rebel movements would recruit followers and future combatants more easily when a large proportion of the population doesn't have access to natural resources such as water, land, or forests. And so environmental changes would threaten security, including the human physical, social, and economic aspects of well-being.

Canadian political scientist Thomas Homer-Dixon distinguishes three categories of conflicts linked to a degradation of the environment.[4] They are ideal types that, in real situations often intermix.

Linked to a scarcity of resources, conflicts in the first category arise either around nonrenewable resources such as oil and mineral deposits, or around renewable resources such as fresh water, fish, and the most productive agricultural land. The availability of these resources becomes rarer when they have been degraded or overexploited, in the case, for example, of a great erosion of agricultural land. This scarcity can also be linked to the size of the demand, due to high demographic growth or to an increase in consumption per inhabitant. The nature of mineral or oil deposits and reserves of fresh water and agricultural land is such that these resources can be seized or controlled physically by one human group at the expense of another. It is in fact possible to extract a large quantity of water from a river for irrigation, thereby causing a lack of water for regions downstream, or to practice intensive fishing in coastal regions of another country thanks to rapid and well-equipped boats, thereby depriving the country in question of a precious resource. The *manu militari* annexation of a fertile agricultural region or oil fields is also an attractive option. These wars of resources can be caused as much by environmental changes as by an inequality in access to the resources under autocratic political regimes, where a privileged minority seizes valuable commodities, such as diamonds in Sierra Leone. The rarer a resource, the greater the temptation for such elites to monopolize it for exploitation, creating a shortage for the rest of society and thus social instability. This syndrome is typical in weak, nondemocratic states, where corruption and bad government reign.

The second category is that of conflicts connected to group identity, which

4. T. F. Homer-Dixon, "On the Threshold: Environmental Changes as Causes of Acute Conflict," *International Security* 16, no. 2, (1991): 76–116.

arise when a large population is displaced following a natural disaster or an ecological degradation. The close contact among different ethnic and social groups on the same territory, in conditions of poverty and dispossession resulting from an environmental disaster, engenders tensions, discrimination, and hostility that can degenerate into violent conflict.

The third category is that of conflicts associated with relative poverty, which arise when a society lags behind in development due to the degradation of the environment. A growing income gap compared with neighboring societies feeds popular discontent. Resentments are all the stronger within the population, particularly in its poorest segment, when the economic deterioration is rapid. When a civil conflict flares up the elite, or those presumed responsible for the economic delay, are then targeted. Such revolts and civil wars appear especially in countries where the distribution of wealth is very unequal and where an elite maintains a social and political status quo in order to preserve its privileges.

THEORY TESTED BY FACTS

Homer-Dixon's neo-Malthusian analysis, which I have just summed up, has all the elegance of simplicity. The facts, however, reveal more complex and multiple causal pathways. Most studies on the connection between environmental change and violent conflicts that integrate social science perspectives show that although it might be a matter of legitimate concern, the alarmist predictions on future environmental conflicts are often exaggerated. In reality, very few recent armed conflicts throughout the world can be unquestionably and directly attributed to environmental changes. The situation that comes closest to this is the series of violent events that occurred in Bangladesh in the 1980s and 1990s, which pitted migrants arriving from the very low altitude plains subject to devastating floods against those living in the hills of Chittagong, the only somewhat elevated region of the country where those fleeing the slow rise of the sea level can converge.

In Rwanda, the 1994 genocide, which claimed more than 1 million victims in three months, has been attributed by some to the scarcity of land and its degradation, following the schema of a Malthusian crisis. In reality, it involved above all a political conflict orchestrated by an elite that played the ethnic division card in an attempt to maintain its power. The massacres had a strong ethnic coloration, but that wasn't all: in fact, many moderate Hutus were also massacred. Granted, the population density is very high in Rwanda and agricultural land rare. All the same, the massacres between the Hutus and the Tutsis had already occurred beginning in 1959, at a time when the population of Rwanda was close to three times smaller than in 1994, just before the genocide. Granted, too, some Rwandans continue to explain that a war was necessary to decrease the excess population in order to survive on the available land resources. That a posteriori justification seems nothing other than a pseudoscientific argument seeking to justify an inexcusable horror.

Eleven percent of the Rwandan population was eliminated during the genocide. With a rate of demographic growth over 3% per year, that is equivalent to three and a half years of demographic growth. To gain such a short respite in demographic growth can in no case explain the large-scale massacre of neighbors by machete. By 1998 the population had returned to its pre-1994 genocide level, and this in spite of the flight of hundreds of thousands of refugees to neighboring countries and the increasing incidence of AIDS. Decreasing the birthrate—on average, a Rwandan woman gives birth to 5 children—is a much milder and sustainable means of controlling population growth. The incontestable need to decrease demographic pressure on land that has become scarce and subject to erosion also involves the development of nonagricultural sectors in cities and rural areas; this creates wealth and eases the demands for land. All developing economies in the world have taken this path.

Granted, these arguments assume a rational behavior by those involved, but those who commit genocide do not operate following calculations based on a demographic and economic model. Environmental issues alone cannot, however, explain the sudden and brutal move to irrationality and hatred, even though it is undeniable that for years Rwanda has endured endless conflicts over land and tensions between neighbors and members of the same family involving agricultural land—uncertainties concerning the means of ensuring the survival of one's family in the face of the degradation of the land and episodic droughts have been profound. But political factors have seen to it that this fragile economic and environmental situation has not been resolved through structural economic and institutional changes, as has occurred in most other countries in the world, but have caused it to be transformed into a wave of madness that unfolded over the country for several weeks.

Globally, the number of armed conflicts between states had reached a peak at the end of the cold war in 1991, then had diminished by a third by 2006. The severity of the conflicts, measured by the number of dead on the battlefields, has also been in decline since World War II. Even in sub-Saharan Africa, a region of the world that has an image of perpetual instability, the number of conflicts diminished by half between 1991 and 2006. And these last fifty years have seen a rapid acceleration of environmental changes. A statistical study on civil wars throughout the world, conducted by the International Peace Research Institute of Oslo, concluded that over the period 1980–92, economic and political factors were much more important than environmental factors in explaining the occurrence of armed conflicts within countries.

THE ROLE OF INSTITUTIONS

Faced with a growing scarcity of natural resources, human societies have an arsenal of possible adaptive responses other than violent conflicts: a modification of the institutions that regulate access to those resources and their use; a more equitable redistribution of wealth; technological innovation; the development of the economic sector of services that engenders wealth while consuming fewer natural resources; the migration and commercial exchanges with neighboring regions in order to compensate for the scarcity of a local resource. A decrease in available natural resources could thus be viewed as a trigger for progress and innovation in a society. In many cases, the degradation of the environment or the scarcity of a natural resource does not engender conflicts, but on the contrary constitutes an important incentive for increased cooperation among states: commercial treaties; multilateral environmental agreements on the climate; biodiversity; bilateral agreements on the sharing of transborder resources to address the pollution of rivers that go through several countries, or air pollution that knows no political borders; or mediation by international organizations such as the UN or through diplomats from neutral nations to reach an agreement.

And so we have a very rich palette of institutional measures that can enable the management of competition and conflicts surrounding natural resources in order to keep them from degenerating into armed violence. Such violence is always only a last-resort solution in case of failure of all other national and international mechanisms of management of tensions that can arise from environmental changes. It is therefore essential that we distinguish between the rare violent conflicts engendered by a struggle for control of natural resources, and nonviolent conflicts, which are most often resolved by diplomacy and negotiation among countries. For example, the disagreements among Europe, the United States, China, and India on how the efforts of those countries in the struggle against climate change should be distributed have never degenerated into violence; they are being resolved through negotiation within diplomatic circles, with a degree of constructive effort among countries of the world that no other international challenge had ever generated before. Environmental conflicts that lead to violence are almost exclusively domestic in nature: regional rebellions, and conflicts between local elites and marginalized social groups.

American sociologist Jack A. Goldstone has shown that slow forms of degradation of the environment—the erosion of land, deforestation, depletion of a nonrenewable resource, water pollution—although they increase poverty and can exacerbate social tensions, are in general not a cause of wars between countries, nor of violent ethnic conflicts, nor of armed uprisings by populations. This category of problems is typically resolved by negotiation and compromise rather than by armed violence. The reason for this is simple: war doesn't resolve them at all; the results of an armed conflict will never be an increase in available resources. The eroded land, the stripped mine, the burned forest, or the polluted groundwater will never be restored by resorting to violence.

In addition, war is a very uneconomical use of public money: the cost of war is much higher than that of redirecting the economy toward alternative resources or of acquiring alternative resources through international trade. According to an estimate that Joseph E. Stiglitz, 2001 Nobel Laureate in Economics, and Linda J. Bilmes published in the *Washington Post* in March 2008, the war in Iraq will have cost the United States $3,000 billion.[5] That money would have easily enabled Americans to invest in the renewable energies necessary to make them independent of imported oil while contributing to a resolution of the global warming problem and reviving their economy with technologies of the future. But the war took place because political motivations and ambitions were combined with geopolitical concerns with respect to oil resources in the Middle East.

Many countries engage in disputes with their neighbors over sharing the water from a large transborder river: Egypt and Sudan over the Nile; India and Bangladesh over the Ganges; Israel and Jordan over the Jordan; Hungary and Slovakia over the Danube, and so on. The word *rival*, moreover, derives from the Latin word *rivalis*, which designates those who use the same river. In these situations, the negotiation of a treaty and recourse to international arbitrage have a much greater chance of leading to a long-term and economically advantageous solution than does war. Investment in technologies for water conservation (in agriculture, in industry, and in private consumption) or even the construction of purification or desalination plants will also cost less, will avoid pressure and international sanctions, and will resolve a conflict more permanently than would a war. The average cost of a week of war against a neighboring country is equivalent to the cost of constructing five water desalination plants. If the water at the center of the litigation is used for agriculture (which is the case for two-thirds

5. L. J. Bilmes and J. E. Stiglitz, "The Iraq War Will Cost Us $3 Trillion, and Much More," *Washington Post*, March 9, 2008.

of the fresh water used throughout the world), importing food products is, here too, less costly than war. Several political scientists assert that during the last few decades, no war has been waged for access to water. That is fortunate, because the UN has identified more than two hundred transborder rivers in the world. For a large proportion of them, disputes between states are not always resolved by treaties of cooperation. The concept of "water war" often appears in our media, but it is almost always presented as a worst-case future scenario.

According to Goldstone, natural disasters that occur quickly, such as floods, fires, earthquakes, or industrial accidents, can be the causes of major political conflicts if the response of the elite is inadequate. The catastrophic industrial pollution in Bhopal in 1984 or in Chernobyl in 1986, as well as the earthquakes of 1985 in Mexico and in 1988 in Armenia, was followed by large-scale political mobilization, which led to a change of regime in all four cases. The cause in each case was less the displacement of the people affected, which might have engendered conflicts, than a revolt by the population faced with the ineffectiveness of its leaders in managing the consequences of the disaster, or the leaders' inability to have anticipated the catastrophe. Whether the disaster could be blamed on the leaders—in the case of crumbling infrastructures or badly planned land development—or bad luck—in the case of a tsunami or a volcanic eruption—the population held the regime in place accountable. Closer to home, the delayed and inappropriate reaction of George W. Bush following the devastating impact of Hurricane Katrina on New Orleans in 2005 contributed greatly to the decline in his popularity. The Chinese authorities had learned their lesson, and reacted quickly and efficiently to the earthquake that affected Hunan Province in southwest China in 2007. Natural or industrial disasters thus offer an opportunity for regimes in power to prove their competence or ineffectiveness, and thus to encourage—or not—political mobilization among the victims.

TOO MANY OR TOO FEW RESOURCES

Many conflicts arise not when there is a scarcity of a natural resource following the degradation of the environment but on the contrary, when there is an abundance of natural resources, especially those whose sale on the international marketplace earns large profits. The substantial income derived from the extraction of oil, diamonds, or valuable tropical timber inspires greed and leads to conflicts to gain control over those resources. In sub-Saharan Africa in particular, the discovery of oil fields is more a curse than good news for the development of the countries concerned. Paul Collier, an Oxford University economist, has defended the thesis according to which the richer a country is in mineral resources, the greater the risk of its having an autocratic government, a lack of investment, civil wars, and bad government. The region of Kivu in the Congo experienced its third civil war in 2008. Between 1998 and 2004, the conflict had already claimed more than 180,000 victims (and not the figure of 4 million to 5 million that is often cited). That region has one of the richest mineral sub-soils in Africa. There is gold, copper, cobalt, uranium, zinc, cassiterite, tin, and many other minerals. The potential for quick wealth that these resources offer attracts more or less violent groups, ready to do anything to ensure a monopoly over their claim. The income earned enables those groups to equip themselves with modern weapons and to attract young recruits into their ranks. Collier has calculated, from 2004 data, that the rate of economic growth of African countries that export oil did not exceed the rate of growth of other countries on the continent. Obviously, in the case of the former, the enormous sums of money earned from oil exports are spent less for economic development than to ensure the wealth and military supremacy of the elite.[6] According to a statistical study,[7] in the Horn of Africa regions with more productive vegetation are at greater risk of armed conflicts than are regions suffering dry conditions. Why indeed would one attack a desert region where a poor population and starving livestock can scarcely survive? Another study shows that the frequency of violent conflicts among tribes of pastoralists increases during years of abundant rain. In these nomadic societies, the principal cause of conflicts is the theft of livestock, which occurs much more frequently in very humid years: vegetation is tall and dense, which offers better camouflage for approaching rival camps; the animals are in better health and thus can better endure a rapid flight over the savannah; and there are more temporary water sources, which increase the animals' rate of survival during the escape.

Is it environmental change that engenders armed conflicts, or war that causes a degradation of the environment? The increase in insecurity and conflicts, whether of political or ethnic origin, forces vulnerable populations to regroup in regions that are less exposed to armed violence. Where human concentrations are high, for example around refugee camps, one can see a degradation of natural resources. By contrast, the resources of abandoned regions are under-exploited. For centuries on the high plateaus of Ethiopia and Eritrea, maintaining sustainable agriculture has been possible only through large-scale land and

6. P. Rowhani and E. F. Lambin, "Climate Variability, Malnutrition, and Armed Conflicts in the Horn of Africa," *Climatic Change* 105, nos. 1–2 (2011): 207–22.

7. P. Meier, D. Bond, and J. Bond, "Environmental Influences on Pastoral Conflict in the Horn of Africa," *Political Geography* 26 (2007): 716–35.

water conservation efforts. These tasks were great consumers of manpower for the application of fertilizer on the fields, the construction of terraces and small dams against erosion, and the accumulation of rocks in erosion ditches to block sediments. During periods of war, the young men most capable of this work are recruited by force into government armies, join the opposition forces, or flee their region to safer places. The resulting abandonment of the conservation work causes an erosion and a degradation of the land. Thus, it is often a lack of security, an important cause of the displacement of populations, which is at the origin of the degradation of the environment, and not the reverse.

THE FIRST CLIMATE CONFLICT?

In northeastern Africa, the Sudan, Ethiopia, Eritrea, and Somalia seem to experience endemic intra- and interstate violence. The climate there is arid, the rain is variable from one year to the next, and land productivity varies greatly from region to region. Following the conflicts of the last few decades in Ethiopia, Eritrea, and then in Somalia, Darfur has been on the front page of the news due to the civil war that has been going on there since 2003. In that region, located to the west of Sudan, tribes of nomadic pastoralists of Arab origin have for millennia lived side by side with the blacker-skinned sedentary farmers, often called "Africans." According to an ancestral regime of land access, in the dry season after a harvest, when the rain is rare in the north, the pastoralists would bring their herds to the agricultural land located farther south. Their camels and goats then fed off the remains of the crops while fertilizing farmers' soil with their excrement. There were many exchanges among these groups, whose activities were very complementary. Since 2003, that cooperation has been transformed into a bloodbath: in 2008, the violence tainted with ethnic cleansing had already claimed 300,000 victims and caused the displacement of 2.7 million people—according to unconfirmed estimates. We can assess these bellicose movements in two quite different ways: is this the first great environmental conflict, which would be a direct consequence of climate change, or is it above all a political conflict?

According to one version, the true source of hostilities was a persistent drought that began in the mid-1980s. In 2007, Ban Ki-moon, the secretary-general of the UN, wrote, "Almost invariably, we discuss Darfur in a convenient military and political shorthand—an ethnic conflict pitting Arab militias against black rebels and farmers. Look to its roots, though, and you discover a more complex dynamic. Amid the diverse social and political causes, the Darfur conflict began as an ecological crisis, arising at least in part from climate change."[8] More directly, Jeffrey Sachs, the renowned economist from Columbia University in New York, wrote, "Darfur, at its core, is a conflict of insufficient rainfall."[9] Al Gore called this civil war a "climate crisis." The government of Sudan subscribed fully to that interpretation which, very conveniently, absolved it from any responsibility: if there had been more rain, the war and the humanitarian disaster that followed would not have occurred. The government of Sudan thus had to manage a local conflict not of its making, caused by climate change. It was the industrialized countries that were truly responsible for the change in rainfall regime, and thus the conflict. The situation in Darfur, moreover, would only be a foreshadowing of what we might expect in the semiarid regions of Africa and elsewhere in the world as a consequence of the climate change to come.

The climate argument starts with the observation that there has been a decrease in rainfall in Darfur during the past forty years. More than 1 million people have perished following famines that have affected Sudan and the Horn of Africa. Water sources have dried up, trees have been cut to feed camels, and agricultural output has decreased. Dunes and sandstorms in the Sahara have invaded pastureland on the edge of the desert, making those lands less productive. Many small tribal conflicts have erupted between farmers and pastoralists, the latter being forced to stay with their herds near cultivated regions and permanent water sources during the rainy season. The two sides have gradually become armed and organized, and animosity has settled in among them. Once their herds of camels were decimated by the drought, some pastoralists converted to small farming. Traditionally not having the right to access the most fertile land, they have had to be content with marginal land, rocky, bare, and dry. During that time, the farmer tribes cultivated much more intensively the richest alluvial land, to which they have access through tradition. Members of pastoral tribes have in one generation passed from the status of proud nomads to that of starving farmers after having lost their herds, the source of their pride and the foundation of their ancestral way of life. The most vindictive among them have fed militias that have burned "African" villages after raping and massacring their inhabitants. The conflict thus would above all be motivated by a competition for access to land and water, a competition made fiercer by the drought.

And yet, careful analysis suggests a much more complex story. The meteorological data for the region show no trace of a drought just before the outbreak

8. *Washington Post*, June 16, 2007.

9. J. Sachs, "Ecology and Political Upheaval." *Scientific American*, June 26, 2006.

of the crisis in 2003. During the thirty years that preceded the conflict, rains had been quite variable—as everywhere in the African Sahel, and that has been true for millennia—but they did not decrease, and the extent of their variations did not increase. Two short but severe droughts did occur in 1984 and 1990. They were not followed by conflicts, but caused a temporary displacement of many people. However, in Darfur, as everywhere in the African Sahel, the pattern of rainfall changed at the beginning of the 1970s, or more than thirty years before the conflict. Whereas the period from 1930 to 1960 was relatively humid for the region, there was less rain after 1971. Then, at the end of the 1990s, there was a higher level of rainfall throughout the African Sahel. The year 2002, which immediately preceded the conflict, experienced more rain than in an average year, and the rain was well distributed throughout Sudan.

In that country, an attentive observer quickly discovers that the classification of pastoralists and farmers—or Arabs and Africans—is reductive: many families combine crop and animal farming on the same territory; during the conflict, various tribes (Arab and African) went from the pro-government camp to that of the rebels, or vice versa, and still lived together peacefully; the expansion of mechanized farming practiced on large agricultural farms caused a scarcity of land, both for farmers and pastoralists. Small conflicts arose around natural resources other than land—abundant resources that engender large profits, such as the water from the Nile or tropical wood; in particular, the discovery of oil caused greed and a struggle for control of that very profitable resource, the principal industrial exploiter of which in Sudan today is China.

The political version of the story of the conflict in Darfur originates in the 1989 coup d'etat in Sudan, which put the National Islamic Front in power—even if a conflict between the Furs and the Arabs had already occurred between 1987 and 1989, claiming thousands of victims. Since then, the policy carried out by Sudanese president Omar al-Bashir, accused in July 2008 of war crimes and crimes against humanity, instituted a discrimination tainted with racism against non-Arab populations. The "African" tribes of Darfur have been economically and politically marginalized. The region has received very little state monetary support in the realms of education, health, transportation, and veterinary services. The British colonists had already neglected this part of Sudan, which is far from everything, a prisoner of its geography. In 1994, the region of Darfur was divided into three states, making its principal non-Arab tribe—the Furs—a minority group within each of those states (in Arabic, *Darfur* means "the home of the Fur"). These regions were then subdivided into emirates, almost all of which were allocated to Arab groups. The power of the traditional tribal chiefs (African and Arab) was weakened, which created influence struggles among tribes.

These conflicts were amplified by the arrival of many Arab groups that came from Chad, who had ties to the Arab tribes from Darfur and sought to occupy new lands. The year 2002 was the prelude to the massacres to come: African villages were burned, camels were stolen by the thousands, and there were several hundred local victims.

Faced with this policy born of an ideology proclaiming Arab supremacy, and in response to the injustice and unequal treatment they were enduring, in 2003 the African tribes of Darfur launched a rebellion against the Sudanese government. The armed movements of these rebels, mostly peasants without military experience but whose families had been victims of the violence, were the Sudan Liberation Army (SLA) and the Justice and Equality Movement. They were promoting not separatism but rather the establishment of development projects in Darfur and a decentralized government.

As for the armed Sudanese forces, they had scarcely recovered from the twenty years of civil war in southern Sudan. Omar al-Bashir thus decided to subcontract control of the region to local militias, the Janjawid, who were armed, financed, and guided by the Sudanese government. Besides the conscripts from the regular army, some of their ranks were made up of former members of the "Islamic legion." This legion was created in the 1980s by Colonel Muammar al-Kadhafi, the president of Lybia, with Tuareg nomads from West Africa and members of the Arab tribes from the Sahel who felt dispossessed of their land and their rights. Darfur had served as their rear base in their war against Chad. After the crushing defeat by the Chad army in 1987, the legion was disbanded. Its former members, uprooted but armed, well trained and fed on the discourse of Arab supremacy promoted in the 1970s by Kadhafi, integrated various militias and rebel movements in the region, including the armed bands of the Janjawid movement. Their marching orders were clear: "Change the demography of Darfur and get rid of the African tribes."

The atrocities that followed are well known: between 700 to 2,000 villages were destroyed, there were systematic rapes and massacres, a deliberately sustained famine occurred, millions of people were displaced, the Western media were held at a distance, and humanitarian organizations were systematically intimidated. The government of Khartoum also succeeded in pitting the various tribes of Darfur—there are 177—against one another, and in dividing the SLA into several factions.

Would the armed conflict have occurred without climate instability and droughts? Probably yes, given the political goals that were behind it. Would it have happened if the National Islamic Front had not taken power by force in Sudan? Undoubtedly not. Drought would certainly have led to a more intense competition for natural resources and forced a displacement of populations, but a legitimate and democratic government would have put peaceful mechanisms in place to resolve the tensions. Notably, it would have organized the distribution of aid and economic investment to benefit all the vulnerable populations. Granted, some local disputes doubtless would have been followed by violent actions. But to claim that the crimes against humanity perpetrated in Darfur were caused mainly by climate change amounts to absolving the Sudanese government from any wrongdoing in its discriminatory policies and its support of the atrocities committed by the Janjawid.

THE SHADOW OF THE MILITARY

If it is true that the threats of environmental conflicts are exaggerated, why is this alarmist prediction so widespread in the media? As every good detective knows, the first question that must be asked is, who benefits from the crime? And then, we must wonder why the military elite of the United States, a country that still refuses to become seriously involved in the global fight against climate change, has now made it its cause célèbre. In 2007, eleven retired American generals and admirals signed a report written by the Center for Naval Analyses Corporation that was widely disseminated in the media; it addressed the threat to US national security posed by climate change. The Institute of Strategic Studies of the United States Army War College has also organized colloquia on climate change. Some have seen this as a recovery operation intended to ensure that the new world challenge of fighting global warming would not result in a decrease in military budgets. Attributing a security and a military dimension to climate change would enable the military to retain the share of public money that is spent profusely on the army and the arms industry. With the end of the cold war and the reservations expressed regarding a military approach to the struggle against terrorism in Afghanistan or Somalia, the assumed future climate conflicts would be one way to maintain an obsession with security and lead a worried public to believe it should maintain or reinforce a military presence—because, without any doubt, the US military will have to intervene in regions of the world severely affected by climate change to resolve these new types of hostilities.

Is this theory somewhat paranoid? It is probably jumping to conclusions, especially as the military is not the only interest group to exaggerate the threat of environmental conflicts. Let's note, however, this paradox: the hypothetical threat that climate change would pose to security would require us to maintain a large military investment rather than use a fraction of that money to solve the problem. In 2007, regular global military spending surpassed $1,100 billion, almost half of which was spent by the United States alone—this figure did not include the cost of the wars in Iraq and Afghanistan, estimated by Linda Bilmes and Joseph Stiglitz at $16 billion per month. The cost of financing in the short term to prevent global climate change from reaching a dangerous level would be on the order of $275 billion a year, or 1% of the gross world product, according to the 2007 estimate of the *Stern Review on the Economics of Climate Change*, an authoritative report on the economic dimension of climate change.[10] To bequeath a healthy environment to our grandchildren would thus cost less than a quarter of the annual expenses of our armies.

In reality, environmental changes present a much lesser threat to the national security of rich countries than they do to the human security of the poorest people. In fact, these changes decrease the options of the poorest populations, as well as their ability to adapt to ensure their survival and well-being. There must be a massive transfer of military budgets to those funding development aid, protection of the environment, and adaptation to climate change in the most vulnerable regions in order to improve the well-being of the most destitute populations.

POLITICS RATHER THAN THE ENVIRONMENT

For the most part, armed conflicts have political causes. Understanding the origin of a conflict requires an understanding of a country's institutions; its degree of stability; the characteristics and motivations of its political leaders and insurgent groups; levels of economic development; the cultural constraints that can impede a transformation of value systems; the ethnic, linguistic, and religious makeup of the society; and the degree of democracy and openness of the country to the outside world. The true roots of armed conflicts associated with a degradation of the environment are the different social groups' very unequal access to natural resources, which is a political factor, and not the overall availability of those resources according to a neo-Malthusian schema.

The concept of environmental conflict is thus a simplification: very few true examples indicate that it is anything but rather alarmist speculation. At the very most, environmental changes are a factor among others likely to exacerbate an

10. *Stern Review on the Economics of Climate Change* (Cambridge: Cambridge University Press, 2007).

already latent conflict stemming from social, ethnic, or political issues. This does not at all diminish the necessity of responding to climatic and other environmental changes—for many other good reasons. Unfortunately, a feeling of insecurity and its negative impact on well-being arise less from objective threats than from the perception that we have of them. Inciting fear vis-à-vis future environmental conflicts leads to a decrease in a sense of security, and therefore of happiness.

ENVIRONMENTAL REFUGEES

This chapter examines another dimension of security: the threat of the forced displacement of populations following environmental changes. A sense of being sheltered from danger, an assurance that over the long term you will be able to enjoy the lifestyle you have constructed, the possessions you have accumulated over time, and the social networks you've created, and thanks to which you have a sense of belonging to a community, are important components of happiness. In the poorest countries, for many people this material and social capital accumulated in the place where they have put down roots is even a matter of survival; belonging to a territory and to a group are the only guarantees they have that their basic needs will be met over the long term.

By contrast, the displaced, refugees, and the uprooted of all kinds have lost this vital connection with their place of origin and, at the same time, with that which provided them with a feeling of security and gave meaning to their lives. This rupture is even more painful when it occurs abruptly and against their will.

Can environmental changes or natural disasters cause entire communities to abandon their homes? Are such changes a significant threat to the well-being of people throughout the world? In this chapter, I will examine cases that might be caused by natural disasters as well as those that could be induced by a slow degradation of the environment.

NATURAL DISASTERS

First of all, we must ask whether natural disasters—earthquakes, volcanic eruptions, tsunamis, floods, droughts, cyclones, forest fires, landslides—are happening more often today than in the past, and whether they are more intense than they once were. The frequency with which such disasters are identified and reported in the media has certainly increased. In 1990, 261 natural disasters were recorded by the Center for Research on the Epidemiology of Disasters of the Catholic University of Louvain in Belgium, as opposed to fewer than 50 per year until the beginning of the 1960s. This figure has grown continuously, reaching 414 in 2007. The number of people affected to varying degrees by these natural catastrophes was 211 million in 2007, with some tens of thousands of victims each year. The frequency with which natural disasters associated with the climate are reported—especially flooding, but also cyclones and hurricanes—has increased more rapidly than has the reporting of earthquakes.

There are three possible reasons for this apparent increase in the frequency of natural disasters: they are occurring more often; their impact on human activity has become increasingly severe, because more and more people and infrastructures occupy vulnerable zones; or progress in the global collection and dissemination of information allows us to be increasingly exposed, almost in real time, to reliable data and images of these disasters. This third reason is incontestable: most of us experienced the earthquake and tsunami that struck the coast of Japan in March 2011 as if it had unfolded on our doorsteps, thanks to amateur videos taken by people who were experiencing the tragedy firsthand. It is also a fact that urban, often poor neighborhoods are built on geological fault lines vulnerable to earthquakes, at the foot of active volcanoes, in flood-prone alluvial plains, or along coasts periodically devastated by hurricanes or tsunamis, thereby considerably increasing the number of victims and the cost of the damage when one of these disasters occurs.

Furthermore, with global warming climate-related natural disasters are becoming more frequent, more intense, and of longer duration. Climatologists call them extreme events when they have occurred in a particular place and during a certain interval of time in only 1 to 10% of the cases during a reference period (in general, between 1961 and 1990). An extremely intense daily or monthly pluviometric event thus corresponds to a quantity of rain higher than that which normally falls in 90 to 99% of the days or months of the reference period. These events can be statistically rare in frequency, amplitude, or duration. Their rarity renders the detection of an increase in their frequency of occurrence particularly difficult. The fact that the global climate time series does not go very far back in the past and is not very homogeneous complicates detection.

Nevertheless, the Intergovernmental Panel on Climate Change (IPCC)—which shared the 2007 Nobel Peace Prize with Al Gore—concluded in its 2007 report that there had been an increase in the frequency of some extreme events. The number of heat waves has been increasing since 1950, with in particular an overall increase in very warm nights—the impact of these heat waves on mortality has been discussed in chapter 6. In midlatitude regions, an increasing proportion of annual rainfall comes from a few highly intense pluvial events; these extremely intense rains, which are increasing more rapidly than average precipitation, cause flooding, which is sometimes exacerbated by the more rapid melting of snow and glaciers in the spring. The frequency of tropical cyclones, as well as hurricanes and typhoons, varies greatly from one year to the next, but the storms have been more intense and have lasted longer since the 1970s, which increases their destructive power; a study published in the journal *Nature* in 2008 shows that the most powerful hurricanes and typhoons—with sustained winds of at least 210 kilometers (131.2 miles) per hour—have become stronger during the past twenty-five years, due to an increase in ocean temperatures. With the rise in sea level and the destruction of mangrove forests that protect the coasts, the destruction caused by cyclones has increased. In southern Eurasia, North Africa, and Canada, the climate has become drier since the 1950s. Between 2002 and 2009, Australia has been facing a severe drought, called the Big Dry, whose economic consequences have been dramatic. These tendencies threaten to intensify in the decades to come. And so the impact of natural disasters is becoming greater, particularly those associated with the climate.

MILLIONS OF DISPLACED

In 2007, British aid organization Christian Aid alarmed the public by predicting that at least 1 billion people would be involuntarily displaced between now and 2050, 300 million of whom would be fleeing destruction caused by climate change and natural disasters, thus adding to the flow of international migrants. In 1995 Norman Myers, a British researcher who had sounded the alarm regarding the many environmental problems in the past few decades, estimated that the number of future environmental refugees between now and 2050 would be from 150 million to 200 million. In 2007, the highly influential *Stern Review*, a report on the economic cost of climate change authored by Sir Nicolas Stern, quoted the figure of 200 million displaced by climate change between now and 2050, and the same year Myers revised his estimate higher, to 250 million.[1] In a speech delivered in February 2009 during an international conference in Cape Town, Stern spoke of the "hundreds of millions, probably billions of people who would have to move if you talk about 4-, 5-, 6- degree increases," while the International Organization for Migration adopted a median estimate of 200 mil-

1. Christian Aid, *Human Tide: The Real Migration Crisis* (London: Christian Aid, 2007).

lion environmental migrants between now and 2050. In an apparently higher bid, in June 2009 Care International spoke of 700 million displaced between now and 2050 as a consequence of global warming.[2]

To help put these figures in perspective, the total number of global international migrants in 2005 was 191 million, and at the end of 2008 there were 42 million people worldwide uprooted by force due to wars and persecution, according to the United Nations High Commissioner for Refugees (UNHCR). In 2007, a study by the United Nations University Institute for Environment and Human Security estimated that by 2100, 100 million to 200 million people will have had to leave their land, which would be either submerged by a rise in sea level or made uninhabitable by other forms of environmental degradation. These estimates seem to make the implicit hypothesis that the countries affected by climate change will not have put any adaptive measures in place. According to the same source, in 2007 20 million people were already displaced due to the erosion of arable lands and pollution of groundwater. In 2008 Javier Solana, the European Union high representative for foreign affairs and security policy, presented a report to the European Council on the issue of climate refugees. He alerted these leaders to the threat of having to face a flow of millions of migrants between now and 2020 who would be fleeing the repercussions of climate change and could threaten the multilateral system of global governance.

How much faith should we put in such alarmist predictions? Those thousands of migrants who every month risk their lives on dangerous boats to reach Spain or Italy from North Africa or Senegal, or who attempt to cross the Mexico-US border in perilous desert conditions: are they fleeing drought and desertification, or are they seeking the economic advantages offered by Europe and the United States? Should we blame environmental changes for these waves of migrants, or rather the failure of development policies in poor countries, bad governance, and the extreme economic inequality that exists throughout the world?

Throughout human history, populations have fled adverse political and economic conditions: religious persecution sent many people onto the roads in the sixteenth and seventeenth centuries throughout Europe; in the twentieth century Nazi Germany and the totalitarian regime of the Soviet Union largely contributed to the flow of refugees; in the last decade in Africa, violent conflicts

2. K. Warner, C. Ehrhart, A. Alex de Sherbinin, S. Adamo, and T. Chai-Onn, *In Search of Shelter: Mapping the Effects of Climate Change on Human Migration and Displacement* (Geneva: CARE International, 2009).

have caused millions of internal refugees and displacements. In recent history, the environment seems to have been the cause of at least one major wave of migration, the one that followed the great famine in Ireland, from 1845 to 1852: around 1 million Irish died of hunger because of the potato blight, since the potato was the basic food source there at the time. There followed the largest emigration of the period: one million survivors immediately set off for the New World, toward Canada or the United States, followed by 3 million more before the end of the nineteenth century. In reality, political factors underlay that migration: the distribution of land was very unequal in Ireland, and the country was under the control of England, which imported a large portion of the Irish grain production, even in the midst of the famine; the English government refused to distribute food aid to the starving farmers and at the same time stopped road construction projects, the only source of additional income for the Irish.

WAVES OF REFUGEES

According to a somewhat formal definition, environmental refugees are individuals who have been forced to leave their traditional place of residence, either temporarily or permanently, due to an important perturbation—of natural or human origin—of their environment that has seriously upset their way or quality of life. Whereas migrants move voluntarily in order to improve their economic status, for example, refugees are forced to move due to external circumstances. In the case of environmental refugees, those circumstances can be of three types: disasters that occur abruptly, an expropriation of their land, or a slow degradation of the quality of their environment.

Disasters can be natural—a volcanic eruption, an earthquake, a cyclone—or caused by humans—water or land pollution following an industrial accident, such as at Chernobyl. The thousands of people displaced from New Orleans after Hurricane Katrina enter into that latter category of refugees, as the impact of the natural disaster was, in their case, amplified by the poor management of the coastal zones' protective infrastructures.

Many environmental refugees are victims of expropriations; their land has been requisitioned for a use that cannot coexist with their traditional way of life. This occurs, for example, with the creation of national parks, the granting of forest land to private timber extraction companies, or the construction of hydroelectric dams. The Three Gorges Dam in China has already caused the displacement of more than 1 million people, and will lead to the further displacement of several more million between now and 2020. The World Bank estimated that

100 million people had been displaced following the environmental impact of development projects in the 1990s.

The impact of wars on the environment also leads to forced migrations, due to the placement of antipersonnel mines in agricultural zones, for example, or the use of defoliant agents such as Agent Orange, employed by the United States during the Vietnam War. The inhabitants of small islands or atolls that will soon be submerged by the rise in sea level also run the risk of being among the first contingents of environmental refugees. Lohachara Island has already been abandoned after being swallowed by the water of the Gulf of Bengal. Tuvalu, an archipelago of the Pacific Ocean that includes nine islands, is at risk of disappearing underwater in the coming decades. It passed a treaty with New Zealand bearing on the migration of workers fleeing the archipelago (in 2008, 3,000 of the 11,600 citizens of Tuvalu already lived in Auckland). The Maldives, an archipelago of 1,190 small islands located in the Indian Ocean, also risk being swallowed by the water, because their average altitude is only 1.5 meters (5 feet) above sea level. At the end of 2008, the new president of this small island paradise decided to invest the profits from tourism into a mutual fund intended to insure the country against climate change. This money will contribute to finding new homes in the event of the forced evacuation of its 300,000 inhabitants. India, Sri Lanka, and Australia have already been targeted as possible destinations. The social, emotional, and logistical cost of such a migration would, however, be enormous. And so the president has proposed another solution, in the hope of being an example for the rest of the world: to become the first country in the world whose net carbon emissions are zero.

More than two hundred communities of Inuits in Alaska, threatened by a retreat of the coasts by 3 meters (10 feet) per year due to coastal erosion, also fall into the category of environmental refugees through the submersion of their land. At the end of the Nile delta, the invasion of seawater into irrigated land constitutes a threat that may eventually necessitate the departure of local populations. In the absence of protective measures, an average rise in sea level by 50 centimeters (1.6 feet) can lead to a loss of more than 10% of the land of Bangladesh. A study published in September 2008 in the journal *Science* concludes that by 2100, the seas will have risen by approximately 80 centimeters (2.6 feet).[3] This recent estimate is more than twice the figures advanced in 2007 by the IPCC, which did not include the accelerated movement of the glaciers toward the sea in Antarctica and Greenland.

3. W. T. Pfeffer, J. T. Harper, and S. O'Neel, "Kinematic Constraints on Glacier Contribution to 21st-century Sea-level Rise." *Science* 321, no. 5894 (2008): 1340–42.

As always, beyond the alarmist figures and talk advanced by some international and nongovernmental organizations, rigorous empirical studies conducted by demographers and geographers reveal a more complex reality. Every decision to migrate is motivated not only by "push" factors linked to the place of origin but also by "pull" factors connected to the place of destination. The concept of environmental refugees places all the weight of the decision to migrate on "push" factors, primarily environmental stress in the region of origin, without considering the power of attraction represented by economic opportunities in the region of destination. In reality, the two types of causes are always blended, which makes it hard to distinguish between purely environmental emigrants and those who are emigrating for strictly economic reasons. Environmental conditions are one of the elements in the general context within which individuals make the decision to migrate. The relationship between environment and migration is always indirect and contextual.

Various surveys and statistical analyses following rigorous protocols have been conducted in Africa and Latin America in order to understand what has motivated specific migrations. The most reliable of these studies are founded on longitudinal field surveys, that is, following families of migrants prospectively over several years. In some countries, we do in fact note more intense migration from densely populated regions or those with low rainfall toward less populated regions with more favorable environmental conditions. All the same, there are also significant flows in the opposite direction. The African Sahel experienced the most severe drought of the twentieth century during the 1970s and 1980s. But detailed surveys taken among people who had migrated within Burkina Faso, whose northern region is part of the semiarid Sahel, revealed that fewer than 10% of them attributed their displacement to a lack of water or food in their place of origin. In neighboring Mali, demographic censuses have shown that migration did not increase during or just after the driest years, in 1983–85. Near the Sahel, in northern Ghana, migrations even decreased during the 1970s and the beginning of the 1980s, at the height of the great drought, and many migrants even returned home due to the economic crisis, political instability, and increase in the price of food in the southern region of the country. During those periods of low rainfall, families made do by engaging in nonagricultural activities that earned them a complementary income, and sent members of their families to stay temporarily with better-off relatives who lived farther away. Pastoralists also began to farm a few parcels of land to compensate for the loss of some of their herds. Furthermore, the money that African immigrants sent from Europe to relatives who remained home significantly increased during times of drought.

Several statistical analyses of the causes of migrations show that sociodemographic factors are more important than environmental factors, except in the extreme cases cited above of islands submerged by the rise in sea level. In addition, the results showing the influence of the environment on migration are sometimes counterintuitive. Sabine Henry, a Belgian geographer, undertook a study of Burkina Faso. She has shown that migrants from regions with favorable environmental conditions are more numerous than those who come from regions where rainfall is low and the land degraded. Emigrating indeed necessitates resources, if only to pay for one's trip and to cover the needs of one's family during one's absence. For the Burkinabé who decided to migrate, the destination was much more often a region with favorable rather than unfavorable environmental conditions. Logically, one migrates less often toward zones affected by drought and land degradation. Statistical analyses also show a great difference between temporary migrations, which are part of a strategy for diversifying sources of income, and long-term migrations. The former, which more frequently take place from regions with little rainfall, do not increase during or just after a drought; they can even decrease. Long-term migrations are associated with neither poor rainfall conditions nor a degradation of the land, but with social and economic factors. They depend on age, level of education, ethnic group, main economic activity, and the proximity of the village of origin to urban centers and transportation infrastructure.

These demographic studies thus do not suggest a purely environmentally motivated migration, including in a region of the world such as the African Sahel, which has been affected by drought and great poverty, and which is often associated with the problem of desertification. Granted, environmental changes do contribute to the "push" factors that lead to a decision to migrate, but only in close relation with the socioeconomic context. In particular, migration is influenced by a person's perception of the constraints of his land of origin and the opportunities at his destination; by the abilities of local societies to adapt in the face of economic and environmental changes, notably through a diversification of economic activities; and by the governmental policies influencing the geographic distribution of public investment and social infrastructures. Furthermore, the attractive economic and cultural factors often play a preponderant role in any decision to migrate. It is more likely that a migration from one rural region to another within the same country is partially influenced by environmental factors, especially if it is temporary, than is a long-term international migration.

Behind the debate on environmental refugees is a hidden, sensitive geopolitical issue: the agreement of rich countries to accept migrants from poor countries who are seeking asylum. Today, environmental refugees still do not have a juridical status, or even an internationally recognized definition. If the status of political refugee, defined by the convention relative to the status of refugees adopted in 1951,[4] were legally extended to include environmental migrants, the consequences for immigration policies in Europe and North America would be considerable. This would de facto imply an obligation to assist refugees pushed onto the roads by changes in the natural environment. Requests for financial compensation for any damage sustained could even be legitimized by the courts if it were established that the cause of the forced displacement was climate change caused principally by greenhouse gas emissions from industrialized countries.

International organizations such as the UNHCR are already overwhelmed by the needs of those who are considered refugees according to the conventional definition; at the end of 2008, 16 million refugees and requests for asylum were recorded throughout the world. According to international law, environmental conditions do not constitute a basis for international protection. If the agency of a rich country responsible for immigration estimates that the motivation of an immigrant to leave his country of origin has been a natural disaster or environmental degradation rather than fear of persecution "for reasons of race, religion, nationality, membership of a particular social group or political opinion,"[5] it does not recognize the status of refugee and thus does not grant asylum.

The agreement reached during the World Conference on Disaster Reduction, held in Kobe, Japan, in 2005, implicitly recognized the duty of nations to protect and aid migrants who were fleeing disasters and degradation of the environment without, however, formalizing that principle into obligations for states. Regional organizations have for a long time enlarged their definition of the status of refugee. Thus, the convention governing the aspects unique to the problems of refugees in Africa, adopted by the Organization for African Unity in 1969, permits the granting of refugee status to any person forced to cross national borders following any disaster caused by humans, whether or not he is threatened by persecution. The UNHCR took a first step toward enlarging its mandate by

---

4. Convention and protocol relating to the status of refugees, United Nations High Commissioner for Refugees, http://www.unhcr.org/3b66c2aa10.html.

5. The 1951 Refugee Convention establishing UNHCR.

intervening, at the request of the United Nations, on behalf of environmental refugees from the tsunami of 2004 in Asia and the earthquake in northern Pakistan in 2005.

Facilitating the fully legal acceptance of populations displaced by environmental degradation by countries having the means will enable the relief of much human suffering. This path is not without pitfalls, however: the governments of countries from which these refugees originate will in some way be exonerated from facing their obligations to avoid degrading their environment and to protect their marginal populations from natural disasters. They could in addition cloak their policies of ethnic cleansing, seizing mineral resources, and discriminating against regional minorities behind "natural" disasters. Furthermore, wouldn't an international recognition of the status of environmental refugee threaten to have a negative impact on the number of migrants who will take their chances every day on overcrowded boats tossed onto the sea or through the scorching heat of the desert of the Mexico-US border? Candidates for migration pay a high price to unscrupulous middlemen for their passage, without realizing that they are often buying their ticket for a death by drowning or heat exhaustion or, if they are luckier, for a forced return to their country after spending months in an overcrowded refugee camp.

Not only is our scientific understanding of the impact of environmental changes on migrations still very incomplete, but we must also continue to look very closely at the geopolitical and juridical consequences of this new phenomenon.

ANOTHER ADVANTAGE FOR THE RICH

Extreme situations such as those of islands and coastal regions that are threatened in the short term by the rise in sea levels are indeed accompanied by their share of environmental refugees. All the same, most migrations are made up of individuals seeking better economic conditions elsewhere. That concern is very legitimate and in no way represents a reason to refuse to accept them. The prosperity of the United States, a country that has been open to migration since it was founded, attests to the cultural and economic dynamism that migrants can contribute. Unfortunately, the alarmist predictions about the hundreds of millions of environmental refugees who will spill out onto the roads of the planet in the coming decades often create a reaction of fear, as if we were facing an invasion of barbarians.

However, as was just shown in the preceding chapter regarding violent conflicts, the influence of the environment on potentially massive displacements of populations is often exaggerated: natural disasters are rarely the cause of long-term international migrations. They lead rather to displacements within the country or to nearby countries, and are often only temporary. The poorest populations that endure a natural disaster do not have the resources to finance a long-distance migration and, a fortiori, one to another continent. The more gradual forms of environmental degradation, such as soil erosion or the degradation of the vegetation cover, are rarely direct and single causes for migration, unless they go beyond a critical threshold, and a community can no longer ensure its subsistence from resources on its land.

As regards health and violent conflicts, it is better not to live in one of the poor countries when it comes to natural disasters and environmental changes. Living in a rich country definitely has its advantages. To ward against the effects of the rise in sea level, the Netherlands will invest 2 billion euros per year in the coming decades, in order to avoid the flooding of its lowest agricultural land. Indeed, 26% of the land of that country is below sea level, and close to two-thirds would be flooded in the absence of antistorm dams and dikes. The Netherlands' current system of protecting the coasts, which is already very sophisticated, will no longer be sufficient in the face of seas that might rise by more than 1 meter (3.3 feet), something that could occur within the next two centuries. The new project includes constructing supplementary dikes and basins to collect the overflow from rivers. It will also require the addition of enormous quantities of sand along the coasts, which will be widened by 30 meters to 1 kilometer (33.3 yards to .6 miles), depending on the location. In all, the cost of this program will be 100 billion euros for this century, or between 0.2 and 0.3% of the country's annual gross national product. The islands and atolls of the Pacific and Bangladesh simply cannot afford such an expense.

To conclude, do environmental changes have a negative impact on your feeling of security by accentuating the threat of being forced to leave your land and thereby threatening your happiness? That depends essentially on the place where you have built your home and live your life. The true driving forces of international migrations and conflicts are the enormous economic inequalities among regions of the world—political instability and bad government in the poorest countries—and not climate change or other forms of degradation of the natural environment.

WHEN POOR COUNTRIES
SET AN EXAMPLE

In this final chapter, we will look at some national policies that have been put in place to reconcile a quest for happiness with a preservation of the integrity of nature. These policies illustrate ways in which we can move from an understanding of the close relationship between happiness and the natural environment, on the one hand, to political action, on the other.

In the media, coverage of the challenges of development and environmental protection always concerns the large countries: China, the United States, India, Russia, Brazil . . . And yet some small, poor countries living in the shadow of the news are undertaking social and environmental experiments that deserve recognition. Three of these national experiments will be looked at here. They involve countries that are rather low on the scale of economic development: Vietnam, Costa Rica, and Bhutan. Each country is on an original course of development that, in every case, aims for an increase in the well-being of its population and for environmental conservation. In particular, in these three countries the tropical forests are gaining ground each year, whereas elsewhere tropical deforestation is pursued with alarming speed. The forests involved are rich in biodiversity, trapping large quantities of atmospheric carbon dioxide, protecting the mountainsides from soil erosion, and regulating the water cycle and flooding in periods of heavy rain. The reforestation observed in these countries thus has multiple ecological benefits on local, regional, and global scales.

This environmental progress is associated with social and economic progress, which puts the countries involved on the road to sustainable development. Whereas Vietnam has pursued a path of economic modernization through a mixture of energetic intervention by the state and a liberalization of the economy, Costa Rica has resolutely played the ecotourism and attractiveness card for well-off foreign vacationers; as for Bhutan, it has opted for a certain isolation vis-à-vis foreign influences and for development centered on the spiritual values of Buddhism. These three very different paths illustrate a plurality of options leading to sustainable development.

LIBERALIZATION AND INTERVENTION IN VIETNAM

In the last few decades, Vietnam has experienced a very high rate of economic growth: 7.5% per year on average between 1991 and 2008. It has recently proved to have one of the most dynamic economies in the world, with remarkable results in the reduction of extreme poverty. In 1986, at the beginning of Communist Party reforms, 70% of Vietnamese families lived below the poverty level. In 2008, that percentage was only 15%. Vietnam succeeded in translating its rapid economic growth into true social progress for its people, improving their level of education, access to infrastructures, and life expectancy. Surveys on satisfaction reveal a population that in general considers itself happy. This country is making significant progress in reaching the Millennium Development Goals promoted by the United Nations in 2000. Nevertheless, economic inequality has not diminished, and the ethnic minorities have benefited less from development than has the rest of the population.

During the 1980s, the Vietnamese government realized that it had lost too much time in wars and ideological conflicts, and that it was imperative that the country be modernized for it to hold a respectable place on the global stage. In Vietnam, development is synonymous with modernization. That concept is seen everywhere, from government policies to individuals' private lives, including the discourse of intellectuals. In the economic realm, state-initiated modernization translated into the creation of a market economy. Many small or medium-sized private companies have been created, and foreign capital has been invested massively in the Vietnamese economy. Liberalization and economic decentralization coexist with a Socialist political regime that exercises an authoritarian paternalism whose roots go deep into fifty years of Communism and two thousand years of Confucianism.

In the environmental realm, rapid economic growth and urbanization have resulted in enormous challenges relating to pollution. In three areas in particular, the situation is getting worse: the treatment of solid waste, water quality, and air pollution in urban settings. There is, however, an environmental domain where Vietnam's progress is impressive: after decades of rapid deforestation of tropical ecosystems rich in biodiversity, the forest cover has been increasing nationally since the beginning of the 1990s. This "forest transition" has been studied by Belgian geographer Patrick Meyfroidt: the forest cover has gone from only 28% of the country in 1990 to 38% in 2005, and it continues to increase. Since the beginning of the 1990s, the rate of reforestation has been twice as rapid as that of deforestation during the 1970s and 1980s.

Let's be clear: primary tropical forests continue to be destroyed today in Vietnam, but the national-scale balance between the decrease in forests in certain regions and the increase in others clearly shows overall reforestation. Some of the land located in the mountains where trees had been cut down has been abandoned, thereby leaving room for a natural regeneration of forests. Furthermore, land parcels located in the plains and deltas have been actively replanted with eucalyptus, acacia, and teak, to respond to the need for wood in the industrial and construction sectors, especially in cities. Each year in Vietnam, this reforestation has absorbed a quantity of atmospheric carbon dioxide (CO_2) through vegetation that is higher than the annual CO_2 emissions in the industrial, domestic, and transportation sectors—thanks to this reforestation, Vietnam has attenuated global climate change very slightly during the last decade. What is remarkable is that this reforestation has not been associated with a decrease in the rural population. On the contrary, in fact, the population density in the highlands that have been replanted has gone at the most from 67 to 73 inhabitants per square kilometer between 1990 and 2000. In rural areas, 87% of the population still lives off agriculture. The growth of the population thus does not always lead to deforestation, and can even be associated with reforestation.

Reforestation policies carried out in Vietnam are an integral part of the development policies in rural regions. The collective farms of the Communist era have been dissolved, the land and farming equipment privatized and the agricultural markets liberalized. This privatization has inspired farmers to invest in a more intensive and profitable agriculture, notably devoting most of their available workforce to growing irrigated rice in the fields located in valley bottoms, and maize and fruit trees at the bases of mountains. Thanks to improvements in seeds and fertilizer, these crops have had high yields. There is a high demand for these products in the agricultural markets. This intensive farming in the valleys has led to a partial abandonment of itinerant slash-and-burn farming, which was traditionally practiced on the slopes of hills and mountains. In this system, which the government considered incompatible with modern agriculture, the land was abandoned for twenty years after several years of cultivation before being farmed again. Slash-and-burn farming is very well adapted to sparsely populated forest regions, but it consumes a lot of space and, on steep slopes, is associated with soil erosion and sometimes with landslides. These mountain regions have thus been abandoned or replanted with forests.

At the level of individual villages, a decentralized land-use zoning policy earmarks certain land for farming (land at the bottom of valleys) and other land for forests. The latter are subdivided into "protective forests," to guard against land erosion on uneven slopes and to regulate the hydrologic cycle, "productive forests," and "special use forests," forests that have a high biological or cultural value. Families have been assigned responsibility for managing protective and productive forests. According to a contract, they are responsible for cultivating the land for the first few years of forest regeneration; then they can extract products from the forest (such as medicinal plants, firewood, game) while respecting certain restrictions. They also receive a modest payment in exchange for their commitment to preserve and protect the forest and, in some cases, for planting trees.

In 1992, after signing the Convention on Biological Diversity, an international treaty for the protection of biodiversity, Vietnam created new protected zones and improved the enforcement of their restricted use. It also instigated large-scale reforestation programs with evocative names: "regreening the bare hills" and "reforestation of 5 million hectares." These policies aimed, among other things, to solidify government control over Vietnamese border territories, which are occupied by marginal ethnic groups still attached to traditional ways of life.

The authorities have many reasons for becoming involved in these political reforms. On the one hand, at a time when the discourse on sustainable development is influencing the practices of international institutions, nongovernmental organizations, and some private companies, projecting an image of an attentive student on the subject attracts public and private funds to support development efforts in the country. On the other hand, associating environmental concerns with a program of economic modernization can facilitate the implementation of reforms that are not always popular. Indeed, every social transition produces its share of winners and losers. Potential losers can be won over by a range of arguments. Thailand, the Philippines, and China experienced catastrophic flooding and landslides at the beginning of the 1990s following deforestation. The human tragedy that ensued convinced the governments of those countries to declare logging bans or at least to restrain the exploitation of forests and to promote reforestation. Vietnam adopted similar measures without, however, having experienced catastrophic flooding in the years preceding reforestation policies. Finally, the Vietnamese leaders realized that the rate of forest exploitation by their businesses was rapidly depleting the stock of valuable wood from their forests. Microhistory has it that in 1992, the prime minister decided to prohibit the exportation of wood after visiting some wood processing factories in his country. For this high-level leader, it had become clear that companies were not respecting their quotas of extraction of valuable tropical wood, and that the country's natural capital was being overexploited.

Does Vietnam offer an example of an environmental policy that should be followed? Let's not decide too quickly . . . In fact, in the 1990s, the Vietnamese government continued to promote the expansion of coffee, tea, and hevea growing in the central region of the country, which led to the destruction of primary forests containing the richest biodiversity. These forests went from covering 384,000 hectares (960,000 acres) in 1990 to 187,000 hectares (467,500 acres) in 2000, a loss of half the forests in ten years. Furthermore, large-scale plantations containing a single, exotic tree species characterized by a rapid growth cycle have no value for biodiversity restoration. Such tree plantations represent around half the reforestation that has occurred since the beginning of the 1990s, the rest being a natural regeneration of forests that contain higher biodiversity. To replace millennial forests, which contain considerable wealth in terms of plant and animal species, with plantings of a single species (eucalyptus, acacia, or pine) is scarcely beneficial to the natural environment. Quantitative statistics on the area of reforestation thus hide a qualitative degradation of forests. It's somewhat as if half the houses in France, along with their antique furniture and associated memories, were replaced with canvas shelters, and housing statistics indicated an improvement in the housing situation.

The land-use zoning policies in rural areas have had negative social impacts on the poorest farmers, notably those from ethnic minorities. They had practiced long-fallow slash-and-burn farming in the highlands for centuries. By losing access to part of that land, they had trouble making ends meet. Some of them also had fields at the bottom of valleys, but not all of them, and so their livelihood was threatened. The allocation of land among a village's inhabitants has not always been equitable: the elite of some communities have benefited from their status to seize the best land. Furthermore, only households with the ability to invest and with the necessary competence have been able to make a profit from managing the forests that have been allocated to them.

Finally, Patrick Meyfroidt's research has shown that following reforestation in Vietnam, the legal and above all illegal importing of wood from neighboring countries—Cambodia, Laos, Indonesia, and Malaysia—has increased considerably.[1] Whereas at the beginning of the 1990s Vietnam imported almost no

---

1. P. Meyfroidt and E. F. Lambin, "Forest Transition in Vietnam and Displacement of Wood Extraction Abroad," *Proceedings of the National Academy of Sciences* 106, no. 38 (2009): 16139–44.

wood, between 1998 and 2000, 1.4 million cubic meters (1.8 million cubic yards) of unprocessed wood was imported illegally each year, in addition to the equivalent of 920,000 cubic meters (1.2 million cubic yards) that entered the country legally in the form of paper paste, cut wood, or wood panels. And this importing has only increased since then: it was five times higher in 2006 than in 2000. A few years after the turning point from net deforestation to net reforestation, the wood processing industry in Vietnam was in full growth. Its production more than tripled between 1990 and 2006, whereas over that same period the volume of wood extracted each year from the country's forests remained unchanged. Vietnam has thus exported its deforestation to neighboring countries, where environmental oversight is weaker. If all the wood processed in Vietnam had been taken from its own forests, the reforestation observed since the beginning of the 1990s would only have been 60% of what has taken place. There would still have been a forest transition, but almost twice as slow.

The strategy has thus been the following: let the neighbors deal with the ecological problems caused by deforestation, and Vietnam will not only profit from transforming raw wood into furniture, but will have the added bonus of developing the image of a country that is restoring its forests. The equivalent of 80% of the wood imported by Vietnam has been re-exported to other countries in the form of luxury items such as furniture. The moral responsibility for exporting forest degradation outside Vietnam is thus shared between this country, whose wood processing industry has prospered, and the final recipients of the wood products, which are distributed throughout the world thanks to the far-reaching channels of international trade. This displacement of the problem and its dispersal over the entire planet unfortunately do nothing to resolve environmental problems on a global scale.

ECOTOURISM IN COSTA RICA

Costa Rica is a small country of 4.3 million inhabitants located in Central America. It is one of the rare countries in the world that does not have an army, a choice made back in 1948. The budgets once allocated to the military have been directed toward education, health care, social protections, and the preservation of nature. One of their former presidents, Oscar Arias, received the Nobel Peace Prize in 1987. The literacy rate there is one of the highest in Latin America. A high proportion of the population in Costa Rica declare themselves to be very satisfied with their lives. More than 95% of the energy consumed by the country comes from renewable sources, thanks to enormous investments in hydroelectric, wind, and geothermal energy. The fauna and flora there are among the richest in the world. Between 4 and 5% of all plant and animal species on the planet exist in Costa Rica, and many species exist only there. Twelve thousand species of plants, including 1,500 different orchids, 850 species of birds, 218 species of reptiles, and 205 species of mammals have been found in Costa Rica. The presence of microclimates, different types of land and topographic features, explains this biological wealth. Furthermore, by forming a "bridge" between North and South America, the Central American isthmus constitutes a forced and narrow passage for species that have moved from the north to the south, and vice versa, over the recent geological history of the earth, enriching the biodiversity of this region as they passed through.

At one time, almost all of Costa Rica was covered with forests, with the exception of a few swamps and high-altitude zones. Before the arrival of the Spanish conquistadors fewer than 2% of the forests had been stripped. Since then, they have been largely destroyed by humans, particularly between 1950 and the mid-1980s. The rate of deforestation there was one of the highest in the world: 3.9% per year, or twenty times higher than the average rate of deforestation in all the tropical regions today. During that period, the forest cover was reduced by half. There was even talk of a "forest striptease" of Costa Rica. In 1984, only 20 to 30% of the territory was still covered with forests. Extensive animal farming was the main cause of this deforestation: legislation stated that a colonist at the forest frontier could obtain a deed for his ranch only if he had cleared at least half the land he would occupy and possessed at least one cow per 5 hectares (12.5 acres). To secure access to land through a legal title of ownership thus required deforesting. Only the least accessible forests, located on the steep mountain slopes, were spared. Between 1985 and 2005, the forest cover was stabilized; it then increased and occupied more than 50% of the territory in 2009.

The expansion of forests that occurred at the end of the 1980s is linked to several factors, both economic and political. On the one hand, land used for farming was reduced and much pastureland abandoned due to a decline in the price of beef on the Central American market—in fact, at the end of the 1970s, Costa Rica had become the fourth-largest provider of beef for the United States (there was talk of a "hamburger connection"), but a few years later, the price of meat decreased by half. Secondary forests spontaneously recolonized former pastureland. On the other hand, the preservation of nature became a mobilizing theme in Costa Rica. Private land was transformed into protected zones by offering the owners financial compensation. Twenty-seven percent of the land today is classified as national parkland or nature reserves, that proportion being one of the highest in the world. Those spaces are used for ecotourism, scientific studies on biodiversity, and bioprospecting in the search for new medicines. They benefit from an effective protection policy and are integrated into a national system of protected zones governed by the Ministry of the Environment and Energy.

In 1996, the government of Costa Rica prohibited the razing of the country's forests. To counter the opposition to this unpopular measure, the government also launched a pioneering program of compensating owners of small and medium-sized properties for services provided by the forests that they protected on their parcels. The principle is simple. Under normal conditions, farmers cut down the trees because farming is profitable, whereas maintaining the forest as it is brings them no income. However, that forest provides the population of a region with important "ecosystem services": regulation of the climate, protection from flooding, protection of the land from erosion and landslides, maintenance of water quality, protection of biodiversity, including many medicinal plants and a rich fauna, and so on. In economic terms, those services are called "positive externalities." The absence of a market for these services explains why, when landowners protect their forests and thus enable those services always to be provided to society, they are usually not reimbursed. The Costa Rican government intervened to fill that gap by creating the first national market assigning economic value to forests.

This new forest policy recognizes four ecosystem services provided by forests: the trapping of atmospheric carbon dioxide through photosynthesis; hydrological services linked to the regulation of the water cycle and to the maintenance of water quality; the protection of biodiversity; and the maintenance of the beauty of the landscape. Within the framework of the Costa Rican program of *pagos por servicios ambientales* (payment for ecosystem services), a mechanism was put in place to compensate landowners who decide not to cut down trees on their land for farming: they receive financial compensation for adopting sustainable forest management practices, the preservation and the natural regeneration of forests, or reforestation activities. A twenty-year contract is established between the forest administration and private owners, according to which the latter give up their rights to alter or destroy their forests and the ecosystem services they provide. This means cultivating not grain or bananas but ecosystem services, which are, of course, a bit more abstract.

To avoid deforestation, it is ideally necessary that the reimbursement be equal to or higher than the income that would be generated by farming the same parcel (that is, the "opportunity cost"). At the beginning of the program, the income per hectare varied from $120 to $45 per year for five years, depending on whether it involved new plantings or existing forests to be managed. Once the five payments were made, the forest was to be preserved for another fifteen years without payment. Such compensation was competitive with intensive cattle farming for meat, but not with more intensive farming. The amount of compensation was increased in 2006, during the presidential campaign.

What is the market, and therefore the source of financing, for these services? The great originality of the program is found in the way in which the Costa Rican government was able to identify the principal beneficiaries of these services, and saw to it that the beneficiaries would finance the services they enjoy. The public administration intervenes only as an intermediary in these transactions.

Thus, trapping carbon emissions in the forests benefits above all the automobile drivers and other consumers of fossil fuel, whose carbon dioxide emissions are partially trapped by the photosynthesis of the vegetation. A 3.5% tax on gasoline and diesel fuel finances this program in part. In addition, the industrialized countries that haven't been able to reach their goals in reducing greenhouse gases as promised in the Kyoto Protocol can buy "certified emission reductions" (or carbon credits) for the carbon trapped in the forests of Costa Rica. The Norwegian Ministry of Foreign Affairs and a consortium of hydroelectric companies in that country have bought compensations at $10 per ton of carbon, a price much lower than the cost of an equivalent decrease in CO_2 emissions in Norway. This country bought $2 million worth of carbon credit from Costa Rica.

Forests also play a role in the natural filtration of water and in soil protection, which guarantees quality drinking water with only a small amount of sediments deposited in the water system. The main beneficiaries are private water distribution companies, which could then invest less in industrial filtration; managers of hydroelectric infrastructures, whose dams fill up with sediment less quickly, thereby prolonging the lifetime of the dam and lowering the cost of production; and farmers who irrigate their fields downstream from the basins whose forest cover has been protected. Financial agreements have been negotiated on an individual basis between private companies in these sectors and landowners. More recently, a tax on water consumption was put in place for forest preservation.

Maintaining rich biodiversity is good for all of humanity, because future generations will benefit from this natural capital, notably from the discovery of new medicines. Payment for that maintenance is assumed by the Global Environment Facility, an independent international organization whose mandate is to finance projects that aim to improve the condition of the global environment. Conservation International, a nongovernmental organization, also contributes to this financing, as a representative of civil society.

Finally, the tourist sector benefits greatly from maintaining the beauty of the landscape. Negotiations seeking to have hotels and rafting companies pay for ecosystem services were not yet concluded by 2007. There is a lot of temptation among the many small, private companies to take advantage of these services without paying for them.

In 2007, close to 10,000 Costa Rican landowners had applied to participate in the program of payment for ecosystem services, and there was a waiting list with three times the applicants that financing allowed, which proved the success of the program. The area involved represented 10% of the country. The program will probably be enlarged, because the country had just obtained financing from the World Bank to triple the area that would benefit from these payments and to add rural development goals to the initial objective of protecting nature.

When an African leader asked him what was key to the success of the program for managing the natural environment, the Costa Rican minister of the environment responded: political stability. Only a state that can plan several decades into the future and is supported by good governance over the long term could launch such a program. And a stable government also benefits from the necessary credibility it has among its citizens.

Thanks to its policies promoting the protection of nature, Costa Rica has developed a solid reputation as a "green country" which, coupled with its image as a small, stable, peaceful, and democratic nation, has contributed to a rapid growth in tourism. Wealthy vacationers—mainly from North America—come to Costa Rica to hike in the tropical forests, scuba dive in the reefs along the white sand beaches, or rest and relax. Since the end of the 1980s, when the country became aware that deforestation was destroying its most precious capital, it has become the world leader in ecotourism. That industry strives for three objectives: educate and entertain visitors, protect the integrity of the natural ecosystems, and respect the local communities while providing them with socioeconomic benefits.

Since Costa Rica has been promoting green and responsible tourism, the number of foreign visitors has increased from 250,000 in 1985 to 1.9 million in 2007, reaching the impressive annual ratio of one tourist per two inhabitants. This sector has become the primary source of foreign capital in the country, exceeding the exportation of bananas at the beginning of the 1990s and attracting investors into the rural areas to construct tourist infrastructures and retirement communities. Many jobs have thus been created. Hotels, restaurants, bed-and-breakfasts, and small amusement parks are sprouting up everywhere. It is not unusual for a hotel to have its own forest preserve offering paths by which guests can explore nature.

The host of several international scientific organizations for the study and the protection of tropical ecosystems, Costa Rica has carefully projected to the outside world an image of being a paradise for naturalists and a destination where a sustainable and responsible tourism is practiced. The true dividends of the country's environmental policies are reaped above all in this sector, which brings in almost $2 billion per year. The danger is that the success of ecotourism will become the cause of its own failure: an ever-growing flood of foreign visitors increases the negative impacts of tourism on the environment and the local culture, and destroys the impression that one is exploring a preserved natural world off the beaten path.

Does Costa Rica offer an example of an environmental policy that should be followed? Whatever the answer, the country is seeing a steady stream of foreign delegations hoping to be inspired by its example. Mexico has launched its own program of payment for ecosystem services, founded on Costa Rican principles. International negotiations to succeed the Kyoto Protocol are developing very similar mechanisms on a global scale that would compensate not only reforestation (which is already the case in the Kyoto Protocol) but also a decrease in the rate of deforestation and forest degradation in a country.

However, there is some criticism of Costa Rica's policies. Payments for the ecosystem services provided by forests benefit primarily the farmers who have the most land and who are already more prosperous, compared with poor households. The latter don't always have an official land title, which was once a condition for participating in the program—but that is no longer the case today. In addition, the poorest citizens often do not have sufficient education, access to information, or the ability to make the initial investment for reforestation necessary for sustainable management of a forest. Some experts also assert that the program of paying for ecosystem services finances landowners who, in most cases, would not have cut down their forests anyway, which some studies confirm. Some landowners have used the money they've received as capital to move into new jobs unrelated to farming (such as driving a taxi in a city, owning a restaurant). Engaging in such different economic activities leads to a partial abandonment of farming and to a natural regeneration of the forest. Program participants generally have more forestland on their parcels than do others. It is possible, however, that they owned more at the outset, which gave them more of an incentive to sign up for the program.

Another criticism concerns the "transaction cost" implied in joining the program of payments for ecosystem services. It is rather high, both for the owner and for the government: one must sign up for the program, provide proof that one is indeed the owner of the land, establish the initial level of forest cover and its evolution over time, propose a plan for managing the forest, and be certified through a visit by a forest inspector. Such costs are justified only for large and midsized landowners (from 300 to 20 hectares [750 to 50 acres]). To lower them, several small landowners often appeal to intermediaries, who carry out administrative processes on the group's behalf. The group thus benefits from economies of scale, but if a member does not respect his contractual obligations, the payment is suspended for the entire group. The cost of the transaction represents around 20% of the payments: 7% is taken by the administration that manages the program, and 12 to 18% is paid by the private owners to the intermediaries. A significant portion of this income therefore does not benefit the owner of the forest that is providing the ecosystem service.

To maintain a high level of biodiversity, it is necessary not only to protect a large area of natural habitat but also to avoid its fragmentation. The exploitation of land by a very large number of small landowners—each of whom decides independently of the others either to preserve his forest or, on the contrary, to cultivate the land—creates countless little forest islands of varying size, from a few hectares to several square kilometers. If these fragments are separated from one another by cultivated land or settlements, species, and therefore genes, cannot circulate freely through the landscape. Despite the expansion of the forest area, such fragmentation therefore remains a threat to biodiversity. National parks and biological reserves themselves are not always connected by ecological corridors, even though the Costa Rican government attempts to recreate connectivity between natural ecosystems.

Future policies will therefore have to better target the spaces that will receive payments for ecosystem services, in order to concentrate them over zones that either will contribute most to biodiversity preservation or are the most threatened. In the first phase of the program, fewer than 10% of the zones benefiting from payments were near an active front of deforestation. This type of program, then, is only successful if it maintains the ability to adjust its criteria over time and to adapt to new circumstances. All the same, Costa Rica has been incontestably innovative in its conception and launching of a flexible program for the protection of nature, one adapted to a system of private ownership and the free market. The Costa Rican program of payments for ecosystem services has not only reinforced local institutions but also led to an environmental awareness in all segments of its population and, from international tourism, throughout the world.

THE MIDDLE ROAD IN BHUTAN

How is your GPH, that is, your gross personal happiness? That is a question being asked in the world of private enterprises, thanks to a very small country isolated from the rest of the world: Bhutan. This Buddhist kingdom is located in the eastern part of the Himalayas, in between two giants: China to the north, and India to the south. Its territory extends from tropical plains at 200 meters (666 feet) in altitude, up to snow-covered peaks at almost 8,000 meters (26,666 feet). Most of the population, fewer than 900,000 inhabitants, is concentrated in a few deep valleys isolated from one another. Although the country has never been invaded, Bhutan has nevertheless long lived with the fear of meeting the same fate as its neighbors, Tibet and Sikkim, two regions that have been integrated by force into China and India respectively. The landscape here is grandiose, with many rivers and forests. Centuries of isolation, a Lamaist Buddhist culture (of Tibetan origin), a low population density, and a very hilly terrain have contributed to the preservation of an intact nature here. The country contains a very rich biodiversity: one notably finds the snow leopard, the Bengal tiger, the red panda, and the golden langur. It is also a true botanical paradise, as seen in one of the old names of the kingdom that means "valleys to the south with medicinal herbs."

Bhutan has great political stability. Its leaders have been visionaries on at least two points. In the 1960s, the preservation and sustainable use of the environment constituted a political priority, which has been pursued consistently since then. This objective was formulated in Bhutan close to thirty years before the 1992 Earth Summit in Rio de Janeiro, which made the concept of sustainable development popular. In another domain, in 1972 the king of Bhutan proclaimed that the ultimate objective of the government would be to promote the happiness of the population. National policies would thus be directed toward a pursuit of "gross national happiness" before aiming for the growth of the gross national product—economists at the best Western universities have only recently come to understand that economic growth does not always go hand in hand with happiness . . .

Policies in Bhutan have always sought a "middle path" between materialism and spirituality. The idea is not, of course, to maximize everyone's happiness at any price—some find their happiness in racist acts or by gratuitously taking the lives of animals—but to cultivate a responsible form of happiness. In this country, the concept of happiness is based on Buddhist values of compassion and harmony. It is a matter not of individual happiness but of a collective enterprise, based on relationships between people. The four pillars of development there are economic independence, the promotion of cultural identity, good government, and the preservation and sustainable use of the environment. No fewer than seventy-two indicators have been defined to measure the evolution of the population's happiness during a national survey that takes place every two years. These indicators measure, for example, the frequency of prayers and meditation, of feelings of jealousy, frustration, calm, and compassion, and also the frequency of suicidal thoughts. About 30% of the national budget is allocated to education and health care, to which everyone has access at no cost. To safeguard against external influences and remain master of its development model, Bhutan has only gradually, and only since the 1970s, opened up to the outside world.

The Bhutanese leaders don't reject modernity, but think that its introduction should be slow and controlled, in order to maintain their cultural identity and avoid the social and environmental ills suffered by countries that have grown too quickly. Even today, tourist visas are granted only parsimoniously, and visits by foreigners are very controlled and costly. The country counts more monks than soldiers. Television and the Internet were introduced only in 1999. The sale of tobacco is forbidden, which makes Bhutan the first completely nonsmoking country. Each year, attacks by the fearsome Asian black bear claim more victims than do traffic accidents. To cross the country from east to west takes four days on narrow mountain roads. By declaration of the king, and against the preferences of a large portion of the population, the country passed from a hereditary monarchy to a parliamentary democracy. The first elections were organized in 2008, after two mock elections whose only goal was to educate the people about the electoral process.

Bhutan's environmental policy is inspired by the Buddhist representation of the human being, which maintains that it is integrated into a complex network of relationships that include all forms of existence. Out of this emerges an ethical imperative for the protection of nature. As in Vietnam and Costa Rica, the forest cover in Bhutan is growing. In 1978, 55% of the territory was covered with natural vegetation, but only 22% with dense primary forests unperturbed by human activity. At that time, there was a great strain on the forests because wood was being extracted for construction or fuel, and because much land was being used for pasture, which prevented a healthy regeneration of forests. In 2005, the area of the country covered by natural vegetation had risen to 81%, and the forests occupied 70% of the land. When that spectacular expansion was taking place, the rate of population growth was 3% per year; 20% of the population lived in an urban setting, and the rest depended on farming, a sector that still represents a third of the country's economy.

Three-fourths of Bhutan's energy consumption is fueled by wood, including almost all its noncommercial energy consumption. Households and small businesses have the right to extract at no charge small quantities of wood, as well as other natural products—mushrooms, crosiers of ferns, medicinal herbs, and several hundred varieties of orchids used in cooking—from the state forests, which form the majority of the forest cover. Farmers depend greatly on the forests: they are used as pastureland for livestock; the dead leaves and the humus serve as beds in stables and, after being mixed with livestock manure, as organic fertilizer for the fields; and the water from streams is intensely exploited for many uses. All this might lead to an overexploitation and degradation of the forests. And yet, they are increasing in surface area.

The reforestation of this country is essentially explained by the environmental policies of the state. Forest legislation requires that forests permanently occupy at least 60% of the national territory. The largest national park in the country was created in 1962. Since then, many regions with low population density have been granted legal status for the protection of nature. In 2005, 27% of the land was formally protected, which is more than double the global average. At the end of the 1960s, a new Forest Department replaced private forest companies and took control of all exploitation of the forests, which were thus nationalized. Pastureland was restricted to certain zones. Goats, which eat young tree saplings, could no longer be left to roam free. On the other hand, herds of yaks, whose ecological impact on the regeneration of pastureland isn't as great, still had access to the land. Tree planting was energetically promoted. A series of acts, policies, and royal decrees for the protection of forests and of nature in general have translated into legislation for the sustainable management of forests and for the protection of biodiversity. These laws have created new institutions encouraging rural communities to participate in forest management. The traditional users of forests were thus granted an official role in the extraction of natural products and the protection of forests.

At the origin of reforestation in Bhutan, one finds above all a development philosophy inspired by the Buddhist vision of the world, one that assigns a high value to nature. This mode of development was put in place through policies promoted by a highly centralized government and through a system of landownership that was essentially state run: very little room is left for private ownership, unlike policies in Vietnam.

Does Bhutan offer an example of an environmental policy to be followed? This development model, which is closely associated with traditional Buddhist values, does have some pitfalls. A portion of the urban youth in the country aspires only to taste the fruits of modernity: jeans, televised programs produced by commercial chains, kung-fu movies and Bollywood, Internet games. A few years ago, during a concert of traditional Bhutanese music, the unfortunate musicians in folkloric attire were greeted by a volley of tomatoes. (Perhaps the youth in that place would have preferred to see Madonna or Britney Spears on-stage wearing a skimpy sequined costume.)

Another problem comes from the difficulty of integrating the Nepalese minority located in the south of the country, the Lhotshampas, who speak another language and are for the most part Hindus. Their tradition is very different from that on which the development model in Bhutan is constructed, that of the Drukpas from Tibet. The Lhotshampas migrated to the subtropical plains of southern Bhutan a long time ago, and have coexisted in peace with the Drukpas for hundreds of years. Waves of more recent Nepalese immigration arrived in the south of the country (and in neighboring Sikkim) between the end of the nineteenth century and the middle of the twentieth. In 1985, when the Lhotshampas organized political demonstrations for more democratic representation, Bhutan promoted the principle of "one nation, one people" to homogenize the social structure, reinforce the national identity, and control foreign influences. At the beginning of the 1990s, 98,000 Lhotshampas who were incapable of proving they had resided in Bhutan before 1958 were forcibly expelled from the country after violent protests. They are still living in refugee camps in eastern Nepal. A large number from ethnic minority groups in Bhutan, 13% of the population, were excluded from the elections of 2008, because they were not considered Bhutanese. The very nature of Bhutan's development model is thus difficult to reconcile with ethnic and religious pluralism, even if the law permits religious freedom and the original Nepalese minority is well integrated into all sectors of the country's economy.

The greatest threat that weighs on Bhutan comes from elsewhere: it is global warming. In fact, the economic development of the country depends in part on the production of hydroelectrical energy; the sale of electricity to India is its main source of revenue. The functioning of dams depends entirely on the melting of the snow deposited each year on the Himalayas and held in the glaciers. But the glaciers are retreating and the accumulation of fresh snow is diminishing with global climate change. The mountain streams that in the spring should cause the turbines of hydroelectric dams to turn full force are becoming less forceful, which threatens long-term electricity production. Even worse, the retreat of the glaciers is accompanied by the formation of glacial lakes, that is, an accumulation of a large volume of water behind the moraines. Under the increasing pressure of water building up from the melting glaciers, these natural dams threaten to collapse, unleashing an eruption of water that would destroy everything located downstream. In addition to the human dwellings in the valleys, the hydroelectric dams would be pulverized. At present, among the 2,700 glacial lakes in Bhutan, 24 are potentially dangerous.

As we have seen in the examples of the three countries presented in this chapter, the paths that lead to sustainable development are quite diverse. The interventionism colored with liberalism in Vietnam, the application of free-market mechanisms to ecology in Costa Rica, and the enlightened leadership inspired by spiritual values in Bhutan have all succeeded in improving both the well-being of the populations and the natural environments of their countries. To leave the beaten paths of standard development models and as a pioneer explore ideas inspired by common sense can often bear fruit. The Costa Rican program of paying for ecosystem services is based on a very simple yet radically new economic idea: paying for services provided by the forests. The decision of the leaders in Bhutan to make the happiness—not the wealth—of its population its primary goal is both elementary and revolutionary compared to prevailing economic dogmas. And yet no model is completely perfect: difficulties and adverse effects appear unexpectedly, even when the new policies are adopted with the best of intentions. Whether problems are exported, as in the case of Vietnam; imported, as in the case of Bhutan; or engendered internally through the implementation of new initiatives, as in Costa Rica, they must be faced by those countries without losing sight of the objectives they have set for themselves. The ability to adjust one's trajectory as one goes along, and to learn from past mistakes, are essential conditions for success.

Each country has its own political and cultural heritage, geographic constraints and advantages, specific economic context, and unique social and political dynamics. Bhutan's isolated location as well as its cultural heritage does not predispose it to adopt the development model followed by Vietnam. Similarly, given the wars for which it was the theater in the second half of the twentieth century, Vietnam did not develop the attractive image for wealthy tourists as Costa Rica and Bhutan have done. "One size fits all" solutions, standardized recommendations of experts with little knowledge of the countries, and universal recipes scarcely contribute to sustainable development. In each particular circumstance, a specific and unique path proves to be the best adapted. We don't need to be convinced that replacing the extraordinary diversity of the culinary arts throughout the world with homogeneous and bland fast food and prepackaged meals does little to encourage happiness. The same is true of policies that aim to reconcile authentic human development with nature: we must preserve a diversity of approaches and solutions. In particular, those that have been presented in this chapter are certainly possible for rich countries, but it will be much more difficult to put them in place in those countries given their institutional complexities and the multitude of actors pursuing different objectives and influencing political processes. One encouraging sign: David Cameron, the new prime minister of the United Kingdom, announced in November 2010 that the British government would start measuring the well-being of its citizens to create a national happiness index.

CONCLUSION

DEBRIEFING

In this book, I have been reflecting on an ecology of happiness, the impact of environmental changes on human well-being, starting from the thesis that humanity has an interest in preserving the integrity of nature, because human happiness is dependent on an interaction with the natural environment. Three components of well-being have been examined: the subjective perception of a happy existence, health, and security.

Many recent studies from various scientific disciplines show that a close contact with nature contributes to the happiness of modern humans through a calming effect and a soothing fascination. Nature is sufficiently vast and rich, and has such polyvalence, that a relationship between humans and nature can be approached from many complementary perspectives: utilitarian, naturalist, ecological, aesthetic, symbolic, humanistic, moral . . . Upon contact with nature, the individual experiences emotional and spiritual satisfaction; he or she also develops a perception of the finality of life through the experience of belonging to the natural world.

Environmental changes often have an unexpected impact on our health. However, the relationship between the environment and human health has proved to be less direct and obvious than what we might simplistically be led to believe. Different environmental factors interact not only among themselves but also with the biological evolution of microbes, socioeconomic changes, and modifications in human behavior. Taken together, the negative effects of environmental changes on health appear to be inevitable: the human species will continue to colonize natural habitats, interact with the animal world, construct cities, develop a globalized society, and extract the greatest good from it, since the benefits of this mode of development for human well-being have up to now been considerable. The challenge consists therefore not in returning to a mode of development that would completely eliminate all these risks—which is neither desirable nor even conceivable—but in controlling those risks and minimizing their emergence, as well as in managing new global threats. This approach does not, of course, rule out the eradication of the most dangerous practices and realities that provide no benefit for human well-being, such as the massive emissions of atmospheric pollutants, including those responsible for global warming; the absence of the most elementary hygiene measures in some human settings as well as in animal farming; the trafficking of exotic animals; the hunting and sale of animals with a high potential for infecting humans, such as primates; and the colonization of the most remote natural environments, notably by unimmunized tourists, who run the risk of contracting local illnesses.

Changes in the natural environment do not contribute significantly to the most extreme forms of insecurity. The environment's influence on threats of armed conflict and involuntary displacement is often exaggerated in the rhetoric of some people. On the contrary, history has shown that a modification of institutions regulating access to and the use of resources, and a more equitable distribution of wealth, in most cases help to avoid worst-case scenarios. One might argue all the same that the role of alarmist predictions is precisely to accelerate institutional initiatives and innovations in order to manage competition and conflicts involving natural resources before they degenerate into armed violence. By contrast, for the most vulnerable populations, human security, or the sense of freedom regarding basic needs and one's ability to adapt in the face of external perturbations, is indeed threatened by environmental changes.

THE HAPPINESS OF SOME CREATES THE UNHAPPINESS OF OTHERS

Poor countries are the most vulnerable in the face of environmental changes, and suffer the most negative consequences of those changes, notably in terms of their health, their ability to pursue their aspirations, and thus their happiness. A profound gap exists between regions of the world that are the principal causes of global environmental changes, that is, the rich countries, and those that endure the most harmful effects of those changes, mainly poor countries. The negative impacts of environmental changes are exported to poor countries in several ways. These include a transfer of the most polluting industries to urban centers within emerging economies, and the installation of intensive farming activities on their peripheries. There is also much exporting of toxic waste to poor countries. In the opposite direction, rich countries engage in the massive importing of raw materials, such as tropical wood, whose extraction from the forests causes great environmental damage. The problem of the globalization of emerging diseases, accompanied by a concentration primarily in rich countries of medical personnel and the means devoted to public health research, also has an inequitable impact among countries. Finally, there is climatic change, whose negative consequences primarily affect marginal rural societies, semiarid ecosystems, and ecosystems of mountains, coastal regions, and islands and atolls at low altitude.

Exporting environmental degradation from rich countries to poor countries is not, of course, the ultimate goal: in reality, it represents a subtle mechanism that enables well-being to be imported into the countries that already enjoy it the most. Because this is indeed what is at issue: behind the international transport of merchandise and more or less toxic products, actual exchanges between regions of the world involve a quest for that immaterial product which is happiness. The concentration of well-being and material comfort in the wealthiest places on the planet, accompanied by a distancing of any harmful by-products, is both the most beneficial and the most perverse effect of globalization, depending on the place where one lives and the degree of moral feeling one has.

The ethical priority humanity must assume, then, is to correct—even to eliminate—those perverse mechanisms that see to it that the almost inevitable cost of economic development and globalization as it affects well-being is essentially borne by the poorest populations of the planet, which benefit least from the current industrial and globalized mode of development. In other words, it should not be only those billion people who live in the poorest countries, whose development stagnates or even retreats, or the many victims of economic growth in emerging economies, who assume the burden.

FIVE CHOICES FOR THE FUTURE

To reconcile a quest for happiness with the preservation of the integrity of nature, humanity must make some choices. Some poor countries have set an example by successfully making difficult decisions, which has required both vision and political courage.

The first fundamental choice is between the promotion of a materialistic system of values and the quest for a more authentic and sustainable happiness. A short time ago, the economic sciences were based on the postulate that those two objectives necessarily went hand in hand. Psychology, sociology, and even economics have shown that this is not the case: economic growth does not necessarily lead to happiness for all. To measure the performance of a society and the policies of its leaders, gross national happiness must be chosen over the gross national product. Either we fill our houses with objects and gadgets, which brings only ephemeral satisfaction, because that happiness evaporates as soon as our comfort level adjusts upward and we become aware that our neighbors have acquired the same goods. Or we adopt a less materialistic system of values, which not only makes us happier in the long term but also would have a smaller ecological footprint, and thus would create more opportunities for an enriching interaction with a preserved nature.

The second choice consists in either promoting an urban culture within which lifestyles are shaped by artifacts and built environments that are substituted for nature, or cultivating our biophilic tendencies and increasing the opportunities for them to be expressed. The domesticated and built environment is certainly reassuring from the feelings of control, order, comfort, and security that it provides. However, recent research has shown that it decreases one's psychological well-being and amplifies stress, does not facilitate a restoration of one's spirit, and contributes no meaning to life: it lessens human existence. The built world is also less resilient, its ability to be restored following a perturbation reduced, because it is homogeneous and contains few mechanisms for reorganization following a shock. By contrast, the rather chaotic diversity of natural forms, which are, granted, unforeseeable and escape our control, evokes the wealth, the complexity, and the mystery of nature; it is a source of meaning and awakens a sense of beauty in humans. Out of that flows nature's power over human happiness and its contribution when we must face any disturbance. Nature also offers enough possibilities so that everyone can find in it that which creates his or her own personal form of happiness.

The third choice involves our interactions with animals. Humans can pursue subjugation and domination by seeing animals only as machines intended to produce marketable goods or as objects meant to promote human all-powerfulness in the realm of leisure activities. Industrial slaughterhouses and hunting are extreme examples of the power over the life and death of animals that humans have granted themselves. By contrast, we can see animals as living beings, certainly different from our species, but who deserve respect, who have rights, and with whom we can establish a mutually beneficial *modus vivendi*. Will we see animals as meat lockers, knowing that intensive farming is a permanent source of zoonoses, or as traveling companions, whose comforting and even therapeutic virtues are undeniable and whose well-being we are responsible for?

The fourth choice involves the way in which we use globalization: will we make it a mode of organization that amplifies social inequalities and whose principal beneficiaries are the greatest—the "free riders," who represent only a small percentage of the global population—and the smallest—microbes—or an extraordinary means of engendering and promoting more sustainable modes of development, information on public health, values and attitudes associated with happiness, and new ideas, all thanks to intercultural exchanges? Depending on the way in which it is regulated and channeled, globalization can be a vector of economic and cultural poverty—or of human progress respectful of the diversity of cultures and ecosystems.

The fifth choice is even more essential, because it is on the moral order. It concerns the fundamental option between an egocentric attitude, one that pursues individual happiness regardless of the cost for others and for nature, and altruism, which makes a feeling of responsibility toward the common good the cornerstone of a mode of development. This means choosing between the perpetuation of great inequality, in that which concerns income as well as the degradation of the natural environment, or social justice following a model of equitable and sustainable development. Short-term egocentrism degrades nature and thus kills individual happiness, especially when it is associated with materialism; it also destroys social justice through an inequitable distribution of the harmful impacts of environmental changes; it therefore diminishes the happiness of all humanity. By contrast, developing a sense of the Other—poor populations, future generations, the animal world, and even nature—is a good investment for human happiness. Indeed, a relationship with nature is one of the components of individual happiness. An altruistic attitude with regard to the natural environment is thus justified from an anthropocentric and even an egocentric point of view—it leads to the long-term interests of humanity and therefore of each individual.

FIVE SITUATIONS WITH DUAL ADVANTAGES

Whereas in most strategic decisions involving the future we must make choices that are beneficial to some things and harmful to others, some situations enable us to win on all counts at the same time—to have our cake and eat it, too, so to speak. The search for a way to reconcile human happiness with the preservation of the integrity of nature includes some of these situations with dual benefits. They represent the goldmines of sustainable development. Forced as we are to make difficult choices in most circumstances, when it is possible not to choose among various priorities but to increase our winnings through the adoption of simple measures, it would be absurd to miss such opportunities. Often, however, there is nevertheless a compromise that must be made.

The benefits of the first decision, to eat less meat, are not dual, but quintuple: this decision is beneficial for the health of the consumer; it easily enables a resolution of the problem of hunger in the world; it decreases the ecological footprint on the land, air, water, and oceans; it facilitates a strict regulation of and control over conditions of industrial farming conducive to the emergence of zoonoses; and it favors a guarantee of animal well-being in the production of animal products. If becoming a vegetarian, let alone a vegan, for some is a difficult decision to put into place on a daily basis, or if the gastronomic pleasure of well-cooked meat remains appealing, a compromise that is within everyone's reach consists in decreasing one's daily ration of meat and promoting processes that certify animal well-being. This is not just a sacrifice in the name of ecology, but in fact a subscription for low-cost health insurance.

The second decision concerns the development of cities and rural areas, and the choice of architectural concepts that promote the presence and frequenting of green spaces. The way in which we occupy the land, not only through urbanization, but also through the accessibility of protected natural zones in the countryside, influence the degree to which we can live in close contact with nature. In cities public parks, roof gardens, and lines and groups of trees break the monotony of the cement, absorb some pollutants, decrease urban temperatures, and offer places to gather. In rural areas, a policy that would be beneficial on several levels would consist in multiplying natural reserves open to the public and in making forests and the countryside more accessible to visitors by reopening old roads, creating cycling paths, and putting into place land policies encouraging private owners to allow hikers looking for a natural experience to cross their land.

In planning offices, hospitals, public buildings, and residential neighborhoods, the virtues of looking out over interior or exterior gardens and the presence of green plants also deserves to be taken into account. There are multiple benefits of such arrangements, whether for human well-being, by offering opportunities for mental restoration, practicing sports and outdoor activities, or for natural ecosystems and their biodiversity, which are all the better protected when the services they provide are formally recognized and appreciated by the public. Nevertheless, engaging in activities in natural settings comes with the risk of contracting zoonoses, in particular those that are transmitted by arthropods. A balance must thus be found between a promotion of those activities and a control of their effects on associated health risks, and on natural ecosystems.

The third decision involves the adoption of soft modes of transport that encourage physical exercise and decrease energy consumption and pollution. The benefits, here, too, are multiple for health and for the environment. The daily practice of getting around on foot or by bicycle is a very effective form of preventive medicine and engenders a sense of well-being. The decrease in concentrations of toxic gasses and fine particles can reduce respiratory and cardiovascular diseases. The regular use of public transportation can offer an opportunity to reinforce social or professional ties in the neighborhood. People opt more readily for moving around on foot, by bike, or on public transportation when their environment is concentrated, there is a favorable mix of urban facilities, attractive and well-maintained sidewalks and cycling paths, and frequent and reliable buses, tramways, and trains. But that is not always enough: driving one's personal car will remain the preferred means of transport for people with high incomes, and so driving should be discouraged through policies of taxation, dissuasive parking fees, and roads reserved for public transportation and for cars carrying several passengers. A compromise must nevertheless be found between a limitation on movement in the name of health and environmental concerns, and the well-being associated with mobility without constraint in the pursuit of leisure activities, a rich social life, or simply for one's profession.

The fourth decision involves an international struggle against climatic change and the destruction of valuable natural ecosystems (tropical and equatorial forests, humid zones, mountain ecosystems . . .) and for the protection of biodiversity. The benefits are of course ecological, but they also reduce the threats of natural disasters, flooding of coastal regions, the emergence of diseases, and of compounded ecological effects, which are difficult to predict. Furthermore, global environmental changes increase the vulnerability of the poorest populations, which are also the least capable of reacting to a change in their precarious living conditions. There is, however, a compromise to be reached between the elimination of all sources of degradation of the environment and the climate, which would have a high economic cost, and adapting to the environmental changes that are considered least dangerous to humanity. In other words, the optimal level of change to nature—that which offers the best balance between the positive and negative effects for well-being—is higher than zero. To regain that point of balance may take several decades.

The fifth decision requires a strengthening of international, regional, and national institutions that promote sustainable development and an equitable distribution of resources among different social groups. The benefits are simultaneously a long-term increase in the living conditions of the poorest of the population, a decrease in the degradation of the natural environment, and a reduction in the threat of conflicts and massive displacements of populations, all of which are related to economic and social inequalities. A certain degree of disparity among regions of the world and of different development trajectories will continue to exist, but the possibility of reaching a decent level of well-being must be offered to all inhabitants of the planet. A better exchange of ideas and information on the many experiments of sustainable development policies carried out in different places in the world, such as those that were described in the preceding chapter, will contribute to thickening the file of possible solutions.

The economic development of the previous centuries has enabled an unprecedented increase in the well-being of humanity. All the same, the threat that this development will have negative effects on well-being when the degradation of nature goes beyond a critical threshold is indeed a reality. And so a delicate balance must be found between the objectives of economic development, the preservation of the natural environment, and an improvement in well-being. The pursuit of happiness for everyone and a preservation of nature must be reconciled. We exercise a high degree of control over factors that would enable the success of such an undertaking, because they are the result of individual and collective choices. The many options for consumption and for economic, dietary, climatic policies, for public health, transportation, urbanism, and the development of the land, can make us more or less happy. The way in which the socioeconomic benefits and the environmental costs of development are distributed over the planet is directly connected to the mode of development adopted by rich countries and the well-being of populations in poor countries. Preserving nature is therefore both in the egocentric interest of each person, but also in the anthropocentric interest of human beings, in the name of the moral reason that connects all members of humanity. These arguments in themselves justify a transition toward sustainable development, in addition to a concern for the common good that transcends our species.

A relationship with nature is a source of realization of self, it gives meaning to life and procures happiness. The yearning to interact with the natural environment is inscribed in human nature. To preserve the natural world and its diversity is thus in the profound interest of individuals and of humanity. A positive perception of nature and its benefits promotes the adoption of behaviors that are in accord with sustainable development. Indeed, respect for the environment is based on an affective connection with nature, and thus contributes to human happiness.

ACKNOWLEDGMENTS

I wish to thank everyone whose input helped
to improve my manuscript, in particular Sophie Bancquart,
Catherine Cornu, Jean-Jacques and Daisy Lambin, and
Patrick Meyfroidt. This book has also benefited from the
participation of many colleagues and doctoral students.
I am indebted, finally, to Teresa Fagan for her careful translation
of the original French text and helpful
editorial suggestions.

BIBLIOGRAPHY

INTRODUCTION

Fischbacher, U., S. Gächter, and E. Fehr. "Are People Conditionally Cooperative? Evidence from a Public Goods Experiment." *Economics Letters* 71, no. 3 (2001): 397–404.

Frumkin, H., and A. J. McMichael. "Climate Change and Public Health: Thinking, Communicating, Acting." *American Journal of Preventive Medicine* 35, no. 5 (2008): 403–10.

Isaac, R. M., and J. M. Walker. "Communication and Free-Riding Behavior: The Voluntary Contribution Mechanism." *Economic Inquiry* 26, no. 4 (1988): 585–608.

Komorita, S. S., and C. D. Parks. "Interpersonal Relations: Mixed-Motive Interaction." *Annual Review of Psychology* 46 (1995): 183–207.

Kurzban, R., and D. Houser. "Experiments Investigating Cooperative Types in Humans: A Complement to Evolutionary Theory and Simulations." *Proceedings of the National Academy of Sciences* 102, no. 5 (2005): 1803–7.

Lambin, E. *The Middle Path: Avoiding Environmental Catastrophe.* Chicago: University of Chicago Press, 2007.

CHAPTER 1

Brereton, F., J. P. Clinch, and S. Ferreira. "Happiness, Geography, and the Environment." *Ecological Economics* 65, no. 2 (2008): 386–96.

Easterlin, R. A. "Explaining Happiness." *Proceedings of the National Academy of Sciences* 100, no. 19 (2003): 11176–83.

Ferrer-i-Carbonell, A., and J. M. Gowdy. "Environmental Degradation and Happiness." *Ecological Economics* 60, no. 3 (2007): 509–16.

Frumkin, H. "Beyond Toxicity: Human Health and the Natural Environment." *American Journal of Preventive Medicine* 20, no. 3 (2001): 234–40.

Frumkin, H., and A. J. McMichael. "Climate Change and Public Health: Thinking, Communicating, Acting." *American Journal of Preventive Medicine* 35, no. 5 (2008): 403–10.

Gowdy, J. "Toward a New Welfare Economics for Sustainability." *Ecological Economics* 53, no. 2 (2005): 211–22.

Hartig, T. "Restorative Effects of Natural Environment Experiences." *Environment and Behavior* 23, no. 1 (January 1991): 3–26.

Hartig, T., G. W. Evans, L. D. Jammer, D. S. Davis, and T. Gärling. "Tracking Restoration in Natural and Urban Field Settings." *Journal of Environmental Psychology* 23, no. 2 (2003): 109–23.

Hartig, T., F. G. Kaiser, and E. Strumse. "Psychological Restoration in Nature as a Source of Motivation for Ecological Behaviour." *Environmental Conservation* 34, no. 4 (2007): 291–99.

Hartig, T., and C. C. Marcus. "Healing Gardens: Places for Nature in Health Care." *Lancet* 368 (2006): S36–S37.

Hinds, J., and P. Sparks. "Engaging with the Natural Environment: The Role of Affective Connection and Identity." *Journal of Environmental Psychology* 28, no. 2 (2008): 109–20.

Inglehart, R., R. Foa, C. Peterson, and C. Welzel. "Development, Freedom, and Rising Happiness." *Perspectives on Psychological Science* 3, no. 4 (2008): 264–85.

Kahneman, D., A. B. Krueger, D. Schkade, N. Schwarz, and A. A. Stone. "Would You Be Happier If You Were Richer? A Focusing Illusion." *Science* 312, no. 5782 (2006): 1908–10.

Kaplan, R. "The Role of Nature in the Context of the Workplace." "Urban Design Research," special issue, *Landscape and Urban Planning* 26, nos. 1–4 (1993): 193–201.

Kaplan, R., and M. E. Austin. "Out in the Country: Sprawl and the Quest for Nature Nearby." *Landscape and Urban Planning* 69, nos. 2–3 (2004): 235–43.

Kaplan, R., and S. Kaplan. *The Experience of Nature: A Psychological Perspective.* Cambridge: Cambridge University Press, 1989.

Kaplan, S. "The Restorative Benefits of Nature: Toward an Integrative Framework." *Journal of Environmental Psychology* 15, no. 3 (1995): 169–82.

Kasser, T. *The High Price of Materialism.* Cambridge, MA: MIT Press, 2002.

Kellert, S. R. "The Biological Basis for Human Values of Nature." In Kellert and Wilson, *The Biophilia Hypothesis,* 42–72.

Kellert, S. R., and E. O. Wilson, eds. *The Biophilia Hypothesis.* Washington, DC: Island Press, 1993.

Layard, R. *Happiness: Lessons from a New Science.* New York: Penguin Books, 2005.

Lévi-Strauss, C. *The Raw and the Cooked.* New York: Harper & Row, 1970.

Mayer, F. S., E. Bruehlman-Senecal, and K. Dolliver. "Why Is Nature Beneficial? The Role of Connectedness to Nature." *Environment and Behavior* 41, no. 5 (September 2009): 607–43.

Rehdanz, K., and D. Maddison. "Climate and Happiness." *Ecological Economics* 52, no. 1 (2005): 111–25.

Stilgoe, J. R. "Gone Barefoot Lately?" *American Journal of Preventive Medicine* 20, no. 3 (2001): 244.

Ulrich, R. S. "Biophilia, Biophobia, and Natural Landscapes." In Kellert and Wilson, *The Biophilia Hypothesis,* 73–137.

———. "View through a Window May Influence Recovery from Surgery." *Science* 224, no. 4647 (1984): 420–21.

Van den Berg, A. E., T. Hartin, and H. Staats. "Preference for Nature in Urbanized Societies: Stress, Restoration, and the Pursuit of Sustainability." *Journal of Social Issues* 63, no. 1 (2007): 79–96.

Van Praag, B. M. S., P. Frijters, and A. Ferrer-i-Carbonell. "The Anatomy of Subjective Well-Being." *Journal of Economic Behavior & Organization* 51, no. 1 (2003): 29–49.

Welsch, H. "Environment and Happiness: Valuation of Air Pollution Using Life Satisfaction Data." *Ecological Economics* 58, no. 4 (2006): 801–13.

———. "Environmental Welfare Analysis: A Life Satisfaction Approach." *Ecological Economics* 62, nos. 3–4 (2007): 544–51.

Wilson, E. O. *Biophilia.* Cambridge, MA: Harvard University Press, 1984.

Zidanšek, A. "Sustainable Development and Happiness in Nations." *Energy* 32, no. 6 (2007): 891–97.

CHAPTER 2

Allen, D. T. "Effects of Dogs on Human Health." *Journal of the American Veterinary Medicine Association* 210, no. 8 (1997): 1136–39.

Anderson, W., C. Reid, and G. Jennings. "Pet Ownership and Risk Factors for Cardiovascular Disease." *Medicine Journal of Australia* 157, no. 5 (1992): 298–301.

Beck, A. M., and N. M. Meyers. "Health Enhancement and Companion Animal Ownership." *Annual Review of Public Health* 17 (1996): 247–57.

Fraser, D. "Toward a Global Perspective on Farm Animal Welfare." *Applied Animal Behaviour Science* 113, no. 4 (2008): 330–39.

Friedmann, E., and S. A. Thomas. "Pet Ownership, Social Support, and One-Year Survival after Acute Myocardial Infarction in the Cardiac Arrhythmia Suppression Trial (CAST)." *American Journal of Cardiology* 76, no. 17 (1995): 1213–17.

Frumkin, H. "Beyond Toxicity: Human Health and the Natural Environment." *American Journal of Preventive Medicine* 20, no. 3 (2001): 234–40.

Frumkin, H., and A. J. McMichael. "Climate Change and Public Health: Thinking, Communicating, Acting." *American Journal of Preventive Medicine*, 35, no. 5 (2008): 403–10.

Grain Briefing. *Seized: The 2008 Land Grab for Food and Financial Security.* 2008. Online at www.grain.org/briefings/?id=212.

Koneswaran, G., and D. Nierenberg. "Global Farm Animal Production and Global Warming: Impacting and Mitigating Climate Change." *Environmental Health Perspectives* 116, no. 5 (2008): 578–82.

Lambin, E. F., H. Geist, and E. Lepers. "Dynamics of Land Use and Cover Change in Tropical Regions." *Annual Review of Environment and Resources* 28 (2003): 205–41.

Leader, S. H., and P. Probst. "The Earth Liberation Front and Environmental Terrorism." *Terrorism and Political Violence* 15, no. 4 (2003): 37–58.

McMichael, A. J., J. W. Powles, C. D. Butler, and R. Uavy. "Food, Livestock Production, Energy, Climate Change, and Health." *Lancet* 370 (2007): 1253–63.

McNeill, J. R., *Something New under the Sun: An Environmental History of the Twentieth-Century World.* New York: W. W. Norton, 2001.

Pollan, M. *The Omnivore's Dilemma: A Natural History of Four Meals.* New York: Penguin, 2006.

Scully, M. *Dominion: The Power of Men, the Suffering of Animals, and the Call to Mercy.* New York: St. Martin's Press, 2003.

Serpell, J. "Beneficial Effects of Pet Ownership on Some Aspects of Human Health and Behaviour." *Journal of the Royal Society of Medicine* 84, no. 12 (1991): 717–20.

Singer, P. *Animal Liberation*. 1975. New York: HarperCollins, 2002.

———. "Animal Liberation at 30." *New York Review of Books* 50, no. 8 (2003).

Sunstein, C. R., and M. C. Nussbaum. *Animal Rights: Current Debates and New Directions*. New York: Oxford University Press, 2004.

Walker, P., P. Rhubart-Berg, S. McKenzie, K. Kelling, and R. S. Lawrence. "Public Health Implications of Meat Production and Consumption." *Public Health Nutrition* 8, no. 4 (2005): 348–56.

CHAPTER 3

Anderson, R. M., and R. M. May. *Infectious Diseases of Humans*. Oxford: Oxford University Press, 1991.

Avila, M., N. Saïd, and D. M. Ojcius. "The Book Reopened on Infectious Diseases." *Microbes and Infection* 10, no. 9 (2008): 942–47.

Corvalan, C., S. Hales, and A. J. McMichael. *Ecosystems and Human Well-Being: Health Synthesis, Millennium Ecosystem Assessment*. Geneva: World Health Organization, 2005.

Costello, A., M. Abbas, A. Allan, S. Ball, S. Bell, et al. "Managing the Health Effects of Climate Change." *Lancet* and University College, London Institute for Global Health Commission. *Lancet* 373 (May 16, 2009): 1693–733.

Diamond, J. *Guns, Germs, and Steel: The Fates of Human Societies*. New York: W. W. Norton, 1999.

Gibbons, A. "Civilization's Cost: The Decline and Fall of Human Health." *Science* 324, no. 5927 (2009): 588–89.

Heeney, J. L. "Zoonotic Viral Diseases and the Frontier of Early Diagnosis, Control and Prevention." *Journal of International Medicine* 260, no. 5 (2006): 399–408.

Hippocrates. *Hippocratic Writings*. Translated by John Chadwick and William Neville Mann. New York: Penguin, 1983.

Jones, K. E., N. G. Patel, M. A. Levy, A. Storeygard, D. Balk, J. L. Gittleman, et al. "Global Trends in Emerging Infectious Diseases." *Nature* 451, no. 7181 (2008): 990–93.

King, D. A., C. Peckham, J. K. Waage, J. Brownli, and M. E. J. Woolhouse. "Infectious Diseases: Preparing for the Future." *Science* 313 (2006): 1392–93.

McMichael, T. *Human Frontiers, Environments and Disease: Past Patterns, Uncertain Futures*. Cambridge, Cambridge University Press, 2001.

McNeil, W. H. *Plagues and Peoples*. New York: Doubleday, Anchor Press, 1997.

Morens, D. M., G. K. Folkers, and A. S. Fauci. "The Challenge of Emerging and Re-emerging Infectious Diseases." *Nature* 430 (2004): 242–49.

Patz, J. A., D. Campbell-Lendrum, T. Holloway, and J. A. Foley. "Impact of Regional Climate Change on Human Health." *Nature* 438, no. 17 (2005): 310–17.

Patz, J. A., and D. E. Norris. "Land Use Change and Human Health." In *Ecosystems and Land-Use Change*, edited by R. DeFries, G. Asner, and R. Houghton, 159–67. Geophysical Monograph Series, no. 153. Washington, DC: American Geophysical Union, 2004.

Prüss-Üstün, A., and C. Corvalán. *Preventing Disease through Healthy Environments: Towards an Estimate of the Environmental Burden of Disease*. Geneva: World Health Organization, 2006.

Snowden, F. M. "Emerging and Reemerging Diseases: A Historical Perspective." "Immunology of Emerging Infections," special issue, *Immunological Reviews* 225, no. 1 (2008): 9–26.

Walters, M. J. *Six Modern Plagues: And How We Are Causing Them.* Washington, DC: Island Press, 2003.

World Health Organization. *Un avenir plus sûr: La sécurité sanitaire mondiale au XXIe siècle.* Annual report. Geneva: WHO, 2007.

Zessin, K. H. "Emerging Diseases: A Global and Biological Perspective." *Journal of Veterinary Medicine B* 53, Suppl. 1 (2006): 7–10.

CHAPTER 4

Alten, B., Kampen, H., and Fontenille, D. "Malaria in Southern Europe: Resurgence from the Past?" In *Emerging Pests and Vector-Borne Diseases in Europe*, edited by W. Takken and B. G. J. Knols, 35–58. Wageningen, the Netherlands: Wageningen Academic Publishers, 2007.

Barbour, A. G., and D. Fish. "The Biological and Social Phenomenon of Lyme Disease." *Science* 260, no. 5114 (1993): 1610–16.

Eisen, L. "Climate Change and Tick-Borne Diseases: A Research Field in Need of Long-Term Empirical Field Studies." *International Journal of Medical Microbiology* 298, Suppl. 1 (September 2008): 12–18.

Enserink, M. "Chikungunya: No Longer a Third World Disease." *Science* 318, no. 5858 (2007): 1860–61.

———. "Tropical Disease Follows Mosquitoes to Europe." *Science* 317, no. 5844 (2007): 1485.

Foley, J. A., R. DeFries, G. P. Asner, C. Barford, G. Bonana, S. R. Carpenter, et al. "Global Consequences of Land Use." *Science* 309, no. 5734 (2005): 570–74.

Gage, K. L., T. R. Burkot, R. J. Eisen, and E. B. Hayes. "Climate and Vector-Borne Diseases." *American Journal of Preventive Medicine* 35, no. 5 (2008): 436–50.

Gubler, D. J. "Resurgent Vector-Borne Diseases as a Global Health Problem." *Emerging Infectious Diseases* 4, no. 3 (1998): 442–50.

Kuhn, K. G., D. H. Campbell-Lendrum, B. Armstrong, and C. R. Davies. "Malaria in Britain: Past, Present, and Future." *Proceedings of the National Academy of Sciences* 100, no. 17 (2003): 9997–10001.

Lafferty, K. D. "The Ecology of Climate Change and Infectious Diseases." *Ecology* 90, no. 4 (2009): 888–900.

Ostfeld, R. S. "Climate Change and the Distribution and Intensity of Infectious Diseases." *Ecology* 90, no. 4 (2009): 903–5.

Ostfeld, R. S., G. E. Glass, and R. Keesing. "Spatial Epidemiology: An Emerging (Or Re-emerging) Discipline." *Trends in Ecology & Evolution* 20, no. 6 (2005): 328–36.

Ostfeld, R., and F. Keesing. "Biodiversity and Disease Risk: The Case of Lyme Disease." *Conservation Biology* 14, no. 3 (2000): 722–28.

Pascual, M., and M. J. Bouma. "Do Rising Temperatures Matter?" *Ecology* 90, no. 4 (2009): 906–12.

Pusre, B. V., P. S. Mellor, D. J. Rogers, A. R. Samuel, P. P. C. Mertens, and M. Baylis. "Climate Change and the Recent Emergence of Bluetongue in Europe." *Nature Reviews: Microbiology* 3 (2005): 171–81.

Randolph, S. E. "The Impact of Tick Ecology on Pathogen Transmission Dynamics." In *Ticks: Biology, Disease and Control*, edited by A. S. Bowman and P. A. Nuttall, 40–72. Cambridge: Cambridge University Press, 2008.

———. "Perspectives on Climate Change Impacts on Infectious Diseases." *Ecology* 90, no. 4 (2009): 927–31.

———. "The Shifting Landscape of Tick-Borne Zoonoses: Tick-Borne Encephalitis and Lyme Borreliosis in Europe." *Philosophical Transactions of the Royal Society London B* 356, no. 1411 (2001): 1045–56.

———. "Tick-Borne Encephalitis Incidence in Central and Eastern Europe: Consequences of Political Transition." *Microbes and Infection* 10, no. 3 (2008): 209–16.

———. "Tick-Borne Encephalitis Virus, Ticks and Humans: Short-Term and Long-Term Dynamics." *Current Opinion in Infectious Diseases* 21, no. 5 (2008): 462–67.

Randolph, S. E., and D. J. Rogers. "Fragile Transmission Cycles of Tick-Borne Encephalitis Virus May Be Disrupted by Predicted Climate Change." *Proceedings of the Royal Society of London B*, no. 267 (2000): 1741–44.

Reiter, P. "Climate Change and Mosquito-Borne Disease." *Environmental Health Perspectives* 109, Suppl. 1 (March 2001): 141–61.

———. "Global Warming and Malaria: Knowing the Horse before Hitching the Cart." *Malaria Journal* 7, Suppl. 1 (December 11, 2008): S3.

Rogers, D. J. "The Dynamics of Vector-Transmitted Diseases in Human Communities." *Philosophical Transactions of the Royal Society of London B*, no. 321 (1988): 513–39.

Rogers, D. J., and S. E. Randolph. "The Global Spread of Malaria in a Future, Warmer World." *Science* 289, no. 5485 (2000): 1763–66.

Smith, D. L., J. Dushoff, and F. E. McKenzie. "The Risk of a Mosquito-Borne Infection in a Heterogeneous Environment." *PLoS Biology* 11, no. 2 (2004): 1957–61.

Stenseth, N. C., N. Samia, H. Viljugrein, K. L. Kausrud, M. Begon, et al. "Plague Dynamics Are Driven by Climate Variation." *Proceedings of the National Academy of Sciences* 103, no. 35 (2006): 13110–15.

Sumilo, D., L. Asokliene, A. Bormane, V. Vasilenko, I. Golovljova, and S. E. Randolph. "Climate Change Cannot Explain the Upsurge of Tick-Borne Encephalitis in the Baltics." *PLoS ONE* 2, no. 6 (2007), doi:10.1371/journal.pone.0000500.

Sumilo, D., A. Bormane, L. Asokliene, V. Vasilenko, I. Golovljova, T. Avsic-Zupanc, Z. Hubalek, and S. E. Randolph. "Effect of Socio-Economic Factors on the Differential Upsurge of Tick-Borne Encephalitis in Central and Eastern Europe." *Reviews in Medical Virology* 18, no. 2 (2008): 81–95.

Vanwambeke, S. O., E. F. Lambin, M. P. Eichhorn, S. Flasse, R. E. Harbach, et al. "Impact of Land-Use Change on Dengue and Malaria in Northern Thailand." *EcoHealth* 4, no. 1 (2004): 37–51.

Wearing, H. J., and P. Rohani. "Ecological and Immunological Determinants of Dengue Epidemics." *Proceedings of the National Academy of Sciences* 103 (2006): 11802–76.

Wilcox, B. A., and R. R. Colwell. "Emerging and Re-emerging Infectious Diseases: Biocomplexity as an Interdisciplinary Paradigm." EcoHealth, no. 2 (2005): 1–14.

CHAPTER 5

Benoiston de Châteauneuf, L. F., J. B. Chevalier, L. Devaux, et al. Rapport sur la marche et les effets du cholera-morbus dans Paris et les communes rurales du département de la Seine. Paris: Imprimerie Royale, 1834.

Johnson, N. P., and J. Mueller. "Updating the Accounts: Global Mortality of the 1918–1920 'Spanish' Influenza Pandemic." Bulletin of the History of Medicine 76, no. 1 (Spring 2002): 105–15.

Krause, R. M. "The Origin of Plagues: Old and New." Science 257, no. 5073 (1992): 1073–78.

Morens, D. M., G. K. Folkers, and A. S. Fauci. "Emerging Infections: A Perpetual Challenge." Lancet: Infectious Diseases, no. 8 (2008): 710–19.

———. "The 1918 Influenza Pandemic: Insights for the 21st Century." Journal of Infectious Diseases 195, no. 7 (2007): 1018–28.

Morens, D. M., and R. J. Littman. "Epidemiology of the 'Plague of Athens' 430–426 BC." Transactions and Proceedings of the American Philological Association 122 (1992): 271–304.

Murphy, V. "Global Killers: Pandemics through History." BBC News, 2005.

Riley, S., C. H. Fraser, C. A. Donnelly, A. C. Ghani, et al. "Transmission Dynamics of the Etiological Agent of SARS in Hong Kong: Impact of Public Health Interventions." Science 300, no. 5627 (2003): 1961–66.

Tatem, A. J., S. I. Hay, and D. J. Rogers. "Global Traffic and Disease Vector Dispersal." Proceedings of the National Academy of Sciences 103, no. 16 (2006): 6242–47.

Thucydides. The History, Writtone by Thucidides the Athenyan, of the Warre: Whiche Was Betwene the Peloponesians and the Athenyans. London: William Tylle, 1550.

Wilson, M. E. "Travel and the Emergence of Infectious Diseases." Emerging Infectious Diseases 1, no. 2 (1995): 39–46.

CHAPTER 6

Brikowski, T. H., Y. Lotan, and M. S. Pearle. "Climate-Related Increase in the Prevalence of Urolithiasis in the United States." Proceedings of the National Academy of Sciences 105, no. 28 (2008): 9841–46.

Chan, C. H., and X. Yao. "Air Pollution in Mega Cities in China." Atmospheric Environment 42, no. 1 (2008): 1–42.

De Hollander, A. E. M., and B. A. M. Staatsen. "Health, Environment, and Quality of Life: An Epidemiological Perspective on Urban Development." Landscape and Urban Planning 65, nos. 1–2 (2003): 53–62.

Dye, C. "Health and Urban Living." Science 319, no. 5864 (2008): 766–69.

Frumkin, H. "Healthy Places: Exploring the Evidence." American Journal of Public Health 93, no. 9 (2003): 1451–56.

———. "The Measure of Place." American Journal of Preventive Medicine 31, no. 6 (2006): 530–32.

Frumkin, H., and A. J. McMichael. "Climate Change and Public Health: Thinking, Communicating, Acting." *American Journal of Preventive Medicine* 35, no. 5 (2008): 403–10.

Godfrey, R., and M. Julien. "Urbanization and Health." *Clinical Medicine* 2, no. 5 (2005): 137–41.

Grimm, N. B., S. H. Faeth, N. E. Golubiewski, and C. L. Redman. "Global Change and the Ecology of Cities." *Science* 319, no. 5864 (2008): 756–60.

Intergovernmental Panel on Climate Change. *Climate Change 2007: The Physical Science Basis.* Contribution of Working Group I to the Fourth Assessment Report of the Intergovernmental Panel on Climate Change [Solomon, S., D. Qin, M. Manning, Z. Chen, M. Marquis, K. B. Averyt, M. Tignor, and H. L. Miller, eds.]. Cambridge: Cambridge University Press, 2007.

Kahn, J., and J. Yardley. "As China Roars, Pollution Reaches Deadly Extremes." *New York Times*, August 26, 2007.

Koppe, C., S. Kovats, J. Gerd, B. Menne, et al. *Heat-Waves: Risks and Responses.* Copenhagen: World Health Organization, 2004.

Luber, G., and M. McGeehin. "Climate Change and Extreme Heat Events." *American Journal of Preventive Medicine* 35, no. 5 (2008): 429–35.

McMichael, A. J., R. E. Woodruff, and S. Hales. "Climate Change and Human Health: Present and Future Risks." *Lancet* 367 (2006): 859–69.

McMichael, T. *Human Frontiers, Environments and Disease: Past Patterns, Uncertain Futures.* Cambridge: Cambridge University Press, 2001.

Normile, D. "China's Living Laboratory in Urbanization." *Science* 319, no. 5864 (2008): 740–43.

Poumadère, M., C. Mays, S. Le Mer, and R. Blong. "The 2003 Heat Wave in France: Dangerous Climate Change Here and Now." *Risk Analysis* 25, no. 6 (2005): 1483–94.

Robine, J.-M., S. L. K. Cheung, S. Le Ropy, H. Van Oyen, C. Griffiths, J.-P. Michel, and F. R. Herrmann. "Death Toll Exceeded 70,000 in Europe during the Summer of 2003." *Comptes Rendus Biologies* 331, no. 2 (2008): 171–78.

Stott, P., D. A. Stone, and M. R. Allen. "Human Contribution to the European Heatwave of 2003." *Nature* 432, no. 2 (2004): 610–13.

Van der Waals, J. F. M. "The Compact City and the Environment: A Review." *Tijdschrift voor Ecoomische en Sociale Geografie* 91, no. 2 (2000): 111–21.

Younger, M., H. Morrow-Almeida, S. M. Vindigni, and A. L. Dannenberg. "The Built Environment, Climate Change, and Health: Opportunities for Co-Benefits." *American Journal of Preventive Medicine* 35, no. 5 (2008): 517–26.

CHAPTER 7

Barnett, J. "Security and Climate Change." *Global Environmental Change* 13, no. 1 (2003): 7–17.

Barnett, J., and W. N. Adger. "Climate Change, Human Security, and Violent Conflict." *Political Geography* 26, no. 6 (2007): 639–55.

Binningsbo, H. M., I. De Soysa, and N. P. Gleditsch. "Green Giant or Straw Man?

Environmental Pressure and Civil Conflict, 1961–1999." *Population and Environment* 28, no. 6 (2007): 337–53.

CNA Corporation. *National Security and the Threat of Climate Change.* Alexandria, VA: CNA Corporation, 2007.

Collier, P. *The Bottom Billion: Why the Poorest Countries Are Failing and What Can Be Done about It.* Oxford: Oxford University Press, 2007.

Diamond, J. *Collapse: How Societies Choose to Fail or Succeed.* New York: Viking, 2004.

Diehl, P., and N. P. Gleditsch, eds. *Environmental Conflict.* Boulder, CO: Westview Press, 2001.

Flint, J., and A. De Waal. *Darfur: A New History of a Long War.* London: Zed, 2008.

Gleditsch, N. P., H. Hegre, and H. P. Wollebaek Toset. "Conflicts in Shared River Basins." In *Water: A Source of Conflict or Cooperation?*, edited by V. Grover, 39–66. Enfield, NH: Science Publishers, 2007.

Goldstone, J. "Demography, Environment, and Security." In *Environmental Conflict*, edited by P. Diehl and N. P. Gleditsch, 84–108. Boulder, CO: Westview Press, 2001.

Hauge, W., and T. Ellingsen. "Causal Pathways to Conflict." In Diehl and Gleditsch, *Environmental Conflict*, 36–57.

Hendrix, C. S., and S. M. Glaser. "Trends and Triggers: Climate, Climate Change, and Civil Conflict in Sub-Saharan Africa." *Political Geography* 26, no. 6 (2007): 695–715.

Homer-Dixon, T. F. "Environmental Scarcities and Violent Conflict: Evidence from Cases." *International Security* 19, no. 1 (1994): 5–40.

———. "On the Threshold: Environmental Changes as Causes of Acute Conflict." *International Security* 16, no. 2 (1991): 76–116.

Kevane, M., and L. Gray. "Darfur: Rainfall and Conflict." *Environmental Research Letters* 3, no. 3 (2008), doi: 10.1088/1748-9326/3/3/034006.

Lonegran, S. C. "Water and Conflicts: Rhetoric and Reality." In Diehl and Gleditsch, *Environmental Conflict*, 109–24.

Malthus, T. *An Essay on the Principle of Population; or, A View of Its Past and Present Effects on Human Happiness.* London: Reeves & Turner, 1878.

Meier, P., D. Bond, and J. Bond. "Environmental Influences on Pastoral Conflict in the Horn of Africa." *Political Geography* 26, no. 6 (2007): 716–35.

Nordas, R., and N. P. Gleditsch. "Climate Change and Conflict." *Political Geography* 26, no. 6 (2007): 627–38.

Ragleigh, C., and H. Urdal. "Climate Change, Environmental Degradation, and Armed Conflict." *Political Geography* 26, no. 6 (2007): 674–94.

Schneider, G., K. Barbieri, and N. P. Gleditsch, eds. *Globalization and Armed Conflict.* Lanham, MD: Rowman & Littlefield, 2003.

Tir, J., and P. F. Diehl. "Demographic Pressure and Interstate Conflict." In Diehl and Gleditsch, *Environmental Conflict*, 58–83.

United Nations Environment Programme. *Sudan: Post-Conflict Environmental Assessment.* Nairobi: UNEP, 2007.

World Commission on Environment and Development. *Our Common Future.* Oxford: Oxford University Press, 1987.

Zhang, D. D., P. Brecke, H. F. Lee, Y.-Q. He, and J. Zhang. "Global Climate Change, War, and Population Decline in Recent Human History." *Proceedings of the National Academy of Sciences* 104, no. 49 (2007): 19214–19.

CHAPTER 8

Afifi, T., and K. Warner. *The Impact of Environmental Degradation on Migration Flows across Countries.* United Nations University Institute for Environment and Human Security, working paper no. 3. Bonn: UNU-EHS, 2007.

Bates, D. "Environmental Refugees? Classifying Human Migrations Caused by Environmental Change." *Population and Environment* 23, no. 5 (2002): 465–77.

Biermann, F., and Boas, I., "Protecting Climate Refugees: The Case for a Global Protocol." *Environment: Science and Policy for Sustainable Development* 50, no. 6 (2008): 8–17.

Center for Research on the Epidemiology of Disasters. *Annual Reports.* Brussels: Université Catholique de Louvain.

Christian Aid. *Human Tide: The Real Migration Crisis.* Report. London: Christian Aid, 2007.

Elsner, J. B., J. P. Kossin, and T. H. Jagger. "The Increasing Intensity of the Strongest Tropical Cyclones." *Nature* 455 (2008): 92–95.

Environmental Change and Forced Migration Scenarios (EACH-FOR). FP6 Research Project of the European Commission. Online at www.each-for.eu.

Gleditsch, N. P., R. Nordas, and I. Salehyan. *Climate Change, Migration, and Conflict.* Coping with Crisis Working Paper Series. New York, International Peace Academy, 2007.

Henry, S., P. Boyle, and E. F. Lambin. "Modelling Inter-Provincial Migration in Burkina Faso, West Africa: The Role of Socio-Demographic and Environmental Factors." *Applied Geography* 23, nos. 2–3 (2003): 115–36.

Henry, S., V. Piché, D. Ouedraogo, and E. F. Lambin. "Environmental Influence on Migration Decisions in Burkina Faso." *Population and Environment* 25, no. 5 (2004): 397–422.

Intergovernmental Panel on Climate Change. *Climate Change 2007: The Physical Science Basis.* Contribution of Working Group I to the Fourth Assessment Report of the Intergovernmental Panel on Climate Change. Cambridge: Cambridge University Press, 2007.

Kibreab, G. "Environmental Causes and Impact of Refugee Movements: A Critique of the Current Debate." *Disasters* 21, no. 1 (1997): 20–38.

Myers, N. "Environmental Refugees in a Globally Warmed World." *Bioscience* 43, no. 11 (1993): 752–61.

Myers, N., and J. Kent. *Environmental Exodus: An Emergent Crisis in the Global Arena.* Washington, DC: Climate Institute, 1995.

Pfeffer, W. T., J. T. Harper, and S. O'Neel. "Kinematic Constraints on Glacier Contributions to 21st-Century Sea-Level Rise." *Science* 321, no. 5894 (2008): 1340–43.

Ramlogan, R. "Environmental Refugees: A Review." *Environmental Conservation* 23, no. 1 (1996): 81–88.

Renaud, F., J. Bogardi, O. Dun, and K. Warner. *Control, Adapt, or Flee: How to Face*

Environmental Migration? InterSections no. 5. Bonn: United Nations University Institute for Environment and Human Security, 2007.

Reuveny, R. "Climate Change-Induced Migration and Violent Conflict." *Political Geography* 26, no. 6 (2007): 656–73.

Solana, J., and B. Ferrero-Waldner. *Climate Change and International Security.* Brussels: European Commission and the Secretary-General/High Representative, 2008.

Stern, N. *Stern Review on the Economics of Climate Change.* London: HM Treasury, 2007.

Warner, K., C. Ehrhart, A. de Sherbinin, S. Adamo, and T. Chai-Onn. *In Search of Shelter: Mapping the Effects of Climate Change on Human Migration and Displacement.* Geneva: CARE International, 2009.

Westing, A. H. "Environmental Refugees: A Growing Category of Displaced Persons." *Environmental Conservation* 19, no. 3 (1992): 201–7.

CHAPTER 9

Castella, J. C., S. Boissau, N. Hai Thanh, and P. Novosad. "Impact of Forest Land Allocation on Land Use in a Mountainous Province of Vietnam." *Land Use Policy* 23, no. 2 (2006): 147–60.

Chomitz, K. M., E. Brenes, and L. Constantino. "Financing Environmental Services: The Costa Rican Experience and Its Implications." *Science of the Total Environment* 240, nos. 1–3 (1999): 157–69.

De Jong, W., D. D. Sam, and T. V. Hung. *Forest Rehabilitation in Vietnam: Histories, Realities, and Future.* Bogor, Indonesia: Center for International Forestry Research, 2006.

Food and Agriculture Organization. *Global Forest Resources Assessment 2005.* FAO Forestry Paper 147. Rome: FAO, 2006.

Kauppi, P. E., J. H. Ausubel, J. Fang, A. S. Mather, R. A. Sedjo, and P. E. Waggoner. "Returning Forests Analyzed with the Forest Identity." *Proceedings of the National Academy of Sciences* 103, no. 46 (2006): 17574–79.

Kerkvliet, B. J., and D. J. Porter, eds. *Vietnam's Rural Transformation.* Boulder, CO: Westview Press, 1995.

Kleinn, C., L. Corrales, and D. Morales. "Forest Area in Costa Rica: A Comparative Study of Tropical Forest Cover Estimates over Time." *Environmental Monitoring and Assessment* 73, no. 1 (2002): 17–40.

Mather, A. S. "Recent Asian Forest Transitions in Relation to Forest-Transition Theory." *International Forestry Review* 9, no. 1 (2007): 491–502.

Meyfroidt, P., and E. F. Lambin. "The Causes of the Reforestation in Vietnam." *Land Use Policy* 25, no. 2 (2008): 182–97.

———. "Forest Transition in Vietnam and Displacement of Deforestation Abroad." *Proceedings of the National Academy of Sciences* 106, no. 38 (2009): 16139–44.

———. "Forest Transition in Vietnam and Its Environmental Impacts." *Global Change Biology* 14, no. 6 (2008): 1319–36.

Pagiola, S. "Payments for Environmental Services in Costa Rica." *Ecological Economics* 65, no. 4 (2008): 712–24.

Raymond, N. "Costa Rica: Du petit pays "démocratique, sain et pacifique" au leader de l'écotourisme et de la protection de l'environnement." *Études caribéennes*, no. 6 (2007).

Sánchez-Azofeita, G. A., C. Harriss, and D. L. Skole. "Deforestation in Costa Rica: A Quantitative Analysis Using Remote Sensing Imagery." *Biotropica* 33, no. 3 (2001): 378–84.

Sánchez-Azofeita, G. A., A. Pfaff, J. A. Robaldino, and J. P. Boomhower. "Costa Rica's Payment for Environmental Services Program: Intention, Implementation, and Impact." *Conservation Biology* 21, no. 5 (2007): 1165–73.

Stem, C., J. P. Lassoie, D. R. Lee, and D. J. Deshler. "How 'Eco' is Ecotourism? A Comparative Case Study of Ecotourism in Costa Rica." *Journal of Sustainable Tourism* 11, no. 4 (2003): 322–47.

Thinley, L. J. Y. "Values and Development: 'Gross National Happiness.'" Keynote speech, Millennium Meeting for Asia and The Pacific, Seoul, 1998.

Uddin, S. N., R. Taplin, and X. Yu. "Energy Environment, and Development in Bhutan." *Renewable and Sustainable Energy Reviews* 11, no. 9 (2007): 2083–103.

CONCLUSION

Frumkin, H., and A. J. McMichael. "Climate Change and Public Health: Thinking, Communicating, Acting." *American Journal of Preventive Medicine* 35, no. 5 (2008): 403–10.